The Pinball Effect

The Pinball Effect

How Renaissance Water Gardens Made the Carburetor Possible—and Other JOURNEYS THROUGH KNOWLEDGE

JAMES BURKE

LITTLE, BROWN AND COMPANY

BOSTON   NEW YORK   TORONTO   LONDON

To Madeline

First Edition

Library of Congress Cataloging-in-Publication Data

Burke, James
    The pinball effect : how renaissance water gardens made the
carburetor possible—and other journeys through knowledge
/ James Burke. — 1st ed.
        p.   cm.
    Includes index.
    ISBN 0-316-11602-5
    1. Inventions — History.   I. Title.
T15.B765   1996
609 — dc20   95-49601

                    10   9   8   7   6   5   4   3   2

                            HAD
            *Published simultaneously in Canada
        by Little, Brown & Company (Canada) Limited*

        Printed in the United States of America

Contents

CONTENTS

Acknowledgments

I should like to express my gratitude to Jay Hornsby, Helen Sharples O'Leary, John O'Leary and Carolyn Doree for their meticulous and rewarding assistance in research.

I should also like to thank Patricia and Archibald (in whose stimulating company most of this was written), and Madeline (in whose stimulating company most of this was *re*written).

How to Use This Book

The Pinball Effect takes twenty different journeys across the great web of change. There are many different ways to read this book, just as there are many different ways to travel on a web. The simplest way is to read from start to finish, in the manner unchanged since the appearance of alphabetic writing thirty-five hundred years ago. Or you can read the book the way your teacher once told you not to. You can do this at many points throughout the book, when the timeline of a particular journey reaches a "gateway" on the web, where it crosses with the timeline of another, different journey. At such a gateway, you'll see the coordinates for the location of that other place.

Using the coordinates you may, if you choose, jump back or forward (through literary subspace) to the other gateway, pick up the new timeline and continue your journey on the web, until you reach yet another gateway, when you may, if you choose, jump once again. The coordinates that identify a gateway appear in the text like this:

... in 1732 that the first of the experiments[14] with the new explosive took place, weeks after he ...

14 29 86

In the text, "experiments[14]" is the site of the fourteenth gateway so far. In the margin, "29 86" is the gateway you'll jump to (the twenty-ninth gateway, located on page 86).

Sometimes there'll be multiple gateways you can jump to, at busy moments in history, where several pathways of change meet. Good luck!

Since there are 447 gateways here which cross, then in one sense, that means this book could conceivably be read at least 447 different ways. Though I don't suggest you try it, doing so would give you a true feel for the pinball effect of the way change happens.

And it happens that way to all of us, all the time. It's happening to you now, though you may not know it yet.

The Pinball Effect

Introduction

We all live on the great, dynamic web of change. It links us to one another and, in some ways, to everything in the past. And in the way that each of us influences the course of events, it also links us to the future we are all busy making, every second. No matter how remote all these links may seem, over space and time, they are real. No person acts without causing change on the web. Each one of us has an effect, somewhere, somewhen. Everybody contributes to the process. In some way, anything we do makes history, because we *are* history. The web is the expression of our existence, and of all those who went before us, and all who will come after us.

Each of the twenty chapter-journeys in this book reveals a different aspect of how the web is constructed and how the extraordinary pinball process of change occurs. Sometimes the simplest act will have cosmic repercussions a hundred years later. On other occasions, the most cataclysmic events will finally lead to ordinary, even humdrum results.

In every case, the journeys presented here follow unexpected paths, because that's how life happens. We strike out on a course only to find it altered by the action of another person, somewhere else in space and time. As a result, the world in which we live today is the end-product of millions of these kinds of serendipitous interactions, happening over thousands of years.

The web can be imagined as a gigantic and ever-growing sphere in space and time, made up of millions of interconnecting, criss-crossing pathways, each one of which is a timeline (a series of points that represent moments when someone or something acted to affect that sequence of events). At the center of this web is the ancient beginning. The surface of the sphere, expanding and growing as every moment goes by, is the modern world.

Extraordinarily, there are some pathways that link the modern world directly to the very ancient, central parts of the web. For example, any time you throw salt over your shoulder, avoid walking under a ladder or cross your fingers, you are repeating an everyday act performed by our Paleolithic ancestors over forty thousand years ago. Everything that surrounds us on the web is, in some way, connected to every other thing, everywhere and everywhen.

This book, for instance, owes its existence to the fact that the German jeweler Johannes Gutenberg got the date wrong one day in the fifteenth century. The lamp you may now be using to read by started life four hundred years ago, thanks to Italian miners whose problems led to learning about the vacuum inside your lightbulb. The discovery of America gave us the science of anthropology. Renaissance water gardens made possible the carburetor. These and many other examples appear in this book, but there are millions more of them making up the web.

The major feature of any journey on the web is the way in which it constantly intrigues and surprises. The book follows the pathways just as they actually happened, putting you in the place of the journey makers of the time. So in almost every case each twist in the path is as unexpected to you as it was to them (and indeed, as it is for us in the modern world, as we take our own journey into the future).

The book has two principal aims. One is to try to conjure up some of the cliff-hanging, detective-story excitement of the pinball way change happens. I also hope the experience of making these journeys will make the readers aware of the ways in which they themselves interact with the world and of the ways in which they are affected by the actions of others. Being aware of change and how it operates is halfway to second-guessing it. And second-guessing change is good for survival and success in life.

The book's other aim is a little more serious. In the coming information age, when multimedia, interactive networking, personal communicators, virtual reality and unlimited bandwidth will all become everyday matters, we will need to think in a different way about knowledge and how it should be used. Supercomputers already manufacture information at speeds no human can match. In the science of quantum chromodynamics, which investigates fundamental particles, there are equations that can be completed, in less than the length of a human lifetime, only by a supercomputer.

In the near future, the rate of change will be so high that for humans to be qualified in a single discipline — defining what they are and what they do throughout their life — will be as outdated as quill and parchment. Knowledge will be changing too fast for that. We will need to reskill ourselves constantly every decade just to keep a job.

As machines increasingly perform what we used to call research and development, and as electronic agents act for us on the networks (delving into the databases to help us with many aspects of our lives, like politics, finance, travel, education and shopping), the new systems will also provide us with unprecedented access to what's going on. Already the Internet provides news faster than live satellite television newslinks. In the near future it will be necessary for everybody to be able to use this information superhighway with the same casual familiarity with which we approach books, newspapers and television. Electronic information sources will become as ubiquitous in our future landscape as the telephone is today.

But these technologies will not enhance our lives if we are not prepared for them. And as the data in our knowledge-bases becomes ever more rapidly obsolete, it will be less important to "know" the data than to be able to access and relate them one to another. Because, as I hope this book shows, it is the interaction between data that causes change. The fundamental mechanism of innovation is the way things come together and connect.

Using the web to watch for and assess the developing patterns of change generated by this kind of interaction will, I think, become a prime social requirement in the near future. The world is too interactive, and the effects of change too wide-ranging, to leave the application of machine-generated innovation to the machine.

If everything any individual does has some effect, then as the massive, high-speed data-processing systems and networks of the future come on-line, the product of their synergy with us will add up to infinitely more than the sum of the parts. The rate of change will accelerate. And, as I hope this book also shows by examples from the past, change may bring social effects for which we are not prepared — unless there is a change in our view of what knowledge is for and how it is generated.

If knowledge is an artifact, and innovation is the result of inter-action on the web, then the way for us to better manage change is to become acquainted with the interactive process. So, in a future world changing too fast for the old-fashioned, specialist approach to education, it may benefit us to require young people to journey the web as a primary learning experience, much in the same way as we taught their ancestors to read books after Gutenberg had in-vented the printing press. Schools might train students to weave their way idiosyncratically through the web, imagining their way to solutions, rather than learning by rote lists of data that will be obsolete before they can use them.

We might even consider changing our definition of intelligence. Instead of judging people by their ability to memorize, to think sequentially and to write good prose, we might measure intelli-gence by the ability to pinball around through knowledge and make imaginative patterns on the web. This book aims to be a first step in that direction. But if nothing else, I hope these journeys on the web are like all other journeys. I hope that they entertain, offer a glimpse of how the world fits together and broaden the mind.

James Burke
London

1 Making Waves

I T doesn't matter where you begin a journey on the great web of change. There is no right place, and no event too humdrum to start from, because one of the fascinating things about the web is the way effects ripple across it. Any action, anywhere, can trigger a chain of events that crosses space and time, to end (perhaps) clear across the world. Anything that happens on the web makes waves.

In this case, the wave maker lived in London in 1906. He was a German hairdresser called Charles Nessler, and his sister hated having to use curlers, which were the only way she could get a wave in her very straight hair. Being a dutiful brother, Nessler kept his eye out for a solution to her problem. One evening he happened to be looking out the window and saw a clothesline which, because of the effect of the dew, had contracted into a wavy shape. So the next day in his salon, he tried re-creating the same effect in his sister's hair and made a discovery that kicked off the multibillion-dollar beauty industry.

Nessler wound his sister's hair onto perforated cardboard tubes, covered it with borax paste, wrapped the tubes in paper to keep out the air and then heated the entire mass for several hours. When he removed the wrappings, the hair emerged beautifully waved — and stayed that way. By 1911, after suffering a lot of ridicule from fellow professionals (and conducting many experiments that went horribly wrong), Nessler had reduced the time his process took to twelve minutes.

The permanent-wave gimmick worked (and made Nessler rich)

because the alkaline borax he was using softened the hair sufficiently for him to be able to remodel it. Then a lengthy period of
heating with hot curlers stiffened the borax, which held the hair in
its new shape. When he had the brilliant idea of calling the process
permanent waving, the technique caught on like wildfire. For the first
time, busy women could come home from work and have time to
set their hair before going out to dinner. Nessler made a fortune
from his "permanent" wave because the effect lasted only a day or
two, after which the process had to be repeated.

Part of the reason why permanent waving was so profitable was
that borax was plentiful and dirt-cheap. Apart from its use in hairdressing, borax was also commonly found in water softeners, plaster, paint, jewelry, china, pottery glazes, electric-light fittings and
antiseptics, as well as being used in metal brazing, refining and
smelting, glassmaking, wood preserving, general cleansing, food
preserving and many other applications.

The industrial process involving borax extraction and processing was simple and inexpensive, and the raw material itself was
plentiful. When Nessler began using it, shiploads of the stuff were
arriving in England from the newly discovered deposits in California, where the mineral appeared in the crust of sand or clay on the
surface of marshes and dry lakes. Turning it into borax was simple:
the crust was mixed with water and boiled up. The reduced solution
was allowed to settle; then it was drawn off into vats, where it
cooled and the borax crystalized out. These crystals were redis-
solved, reboiled and allowed once more to crystalize. This pure
form of crystal was then ground into a fine powder, and the borax
was ready for use.

The California borax had been discovered in Death Valley, some
parts of which are 282 feet below sea level, making it the lowest
place in the United States. In consequence, the valley floor is the
final area of concentration for an immense drainage system, and
large parts of the floor are covered with deep deposits of salt and
nitrate of soda, as well as borax. When the borax was discovered,
there was serious doubt as to whether the deposits could be worked
at all, because of the appalling climatic conditions. In the early
Death Valley mining camps, furniture split and warped within days
of arrival, water barrels left empty would lose their hoops within
an hour, the washed end of a blanket would be dry before the other

end came out of the tub, eggs would roast in the sand and it was said that a man who went without water for an hour would go insane. Death Valley was so named because many of the early settlers trying to pass through it on their way to the West Coast simply didn't survive the experience.

It was in 1841 that these immigrants started to arrive in California in any significant numbers, and those who had made it through Death Valley often stopped at a fort called New Helvetia, situated near what is now Sacramento and run by one of history's greater con men, Johann Sutter. Sutter was quite a piece of work. He had begun life as a store clerk in Berne, Switzerland; then he married, opened his own store and went broke; and in 1834 he skipped the country for New York, leaving his wife and five children destitute. After wandering round America, Canada, Russian Alaska and Hawaii, he finally arrived in California in 1839, along with a few dubious hangers-on he had picked up on the way. By this time he was calling himself "Captain" Sutter and claiming to have been a member of the Pope's Swiss Guard, a French royal guard, a fellow student at university with the French emperor, a Swiss army officer, unmarried and financially solvent (none of which was true).

Sutter's first idea was to set up a colony, but he decided instead to strike out for the wilderness (once he was informed by the local authorities that if the colony ever broke up, its members would share Sutter's claim to colony property). After traveling up the American River as far as he could, Sutter started building a small settlement. For two years it was a matter of touch-and-go whether the place would survive, but by 1841 a fort was in place, manned by Sutter's employees wearing colorful uniforms bought second-hand from nearby Russians when they vacated their settlements along the California coast.

By 1848 Fort Sutter, as it was now called, had proper walls, cannons and brick buildings and was a regular trading post and stopover for settlers on their way into the territory. Sutter was finally beginning to make money when he made the fateful decision to improve his financial position even more with a sawmill that would provide settlers with timber. He chose a site about fifty miles upriver from the fort.

Then, on the dark and dismal evening of January 28, 1848, as rain lashed across the fort, the millwright, James Marshall, arrived at Sutter's quarters, soaked to the skin and breathless with excitement. The cause of his condition was a piece of yellow rock, wrapped in a cloth, which he had brought from the sawmill. Sutter locked the door, and both men carried out various tests on the rock, following instructions from Sutter's encyclopedia. These tests revealed Marshall's find to be a nugget of pure gold. Sutter and Marshall were about to trigger the California gold rush.[1]

1 58 *44*

1 153 *122*

In 1848, before the news of this discovery leaked out, no more than four hundred newcomers had arrived in California from the East. The following year the number had rocketed to ninety thousand. Almost overnight California went from Catholic to Protestant, from Latino to Anglo-Saxon and from an ox-cart economy to one in which, for a price, you could buy anything you wanted — no matter where it came from or how difficult it was to find. The great American rags-to-riches myth was born, and with it the dream of the golden West. Cities across America emptied, as their populations headed for the goldfields after reading newspaper reports of inexperienced prospectors making as much as $15,000 in a single day.

There were many other ways to get rich in the goldfields apart from mining. More than one hundred thousand miners needed food, clothes and housing, so shopkeepers, farmers, cooks, unmarried girls and entrepreneurs of all kinds flooded into California to make their fortune. One woman said she'd made $18,000 selling pies (at a time when $500 would support a family for several years). Two acres of onions reportedly went for $2,000. Oregon apples sold for $1.50 each. The gold rush newcomers came from all over the world, a fact reflected in the names of the mining communities: French Creek, Cape Cod Bar, Georgia Slide, Chinese Camp, German Bar, Dixie Valley, Dutch Flat.

Early on it became clear that the first arrivals would very likely make the most money because they would get there before the easiest pickings ran out. But the cross-country journey by covered wagon (suffering some of the most difficult terrain on earth, Indian attacks and disease) could take six or seven months. For the serious fortune hunter, the fastest way to go was round the Horn on board a Yankee clipper, the fastest ship in the world.

The clipper was a radically new kind of ship, built mostly by American yards. At a time when cargo vessels were broad and slow, the clipper's new design sported a long, slim hull and a lean, sharp prow because the ship was built to carry a lightweight cargo at maximum speed. Perhaps the most extraordinary thing about the clipper was the rigging. When the first "extreme" clipper took to the water in 1845, nobody had ever seen a greater spread of sail.

The clipper builders added extra masts to carry more canvas, splitting sails to make it easier to raise or strike them in winds of up to hurricane strength. It was the boast of the clipper captains that they never hauled in sail, even in the worst weather.

The clipper was originally built not for the California run, but to go from England to China, thanks to the British addiction to tea.[2] In the mid–nineteenth century Britain was consuming the staggering total of eighty million pounds of tea a year, and there were thirty thousand wholesale and retail tea dealers in London alone. So the U.K. demand for China tea represented a tremendous business opportunity. By 1850 the new Yankee clippers were beating the old-style British ships, racing back to London through all kinds of weather, carrying the best of the tea crop and naming their own price for it.

The reason the venture paid so handsomely was that the British tea drinker believed that tea lost its fragrance during a long voyage home. So the youngest, most flavorsome leaves of each new crop went to the ships with a reputation for speed. The clippers earned their reputation during some of the most exciting races in maritime history. On one occasion, in 1866, in a five-ship race, the two lead ships arrived in London less than an hour apart — after a journey of 101 days.

It was ironic that the great Yankee clippers were given the chance in the first place to beat British maritime trade at its own game only because of an American fungus that had crossed the Atlantic in 1846 and caused widespread death and destruction in Ireland.

At that time the absentee English landlords, who owned most of the country, rented their property at a fixed price to middlemen. These agents boosted their profits by subdividing land parcels into the minimum size that would support one family, and then charging extortionate rents. They also imposed contracts requiring that any improvement made by the tenant pass to the landlord at the end of a lease.

These were the years when the Irish population was rising at an unprecedented rate due to the diet, which relied overwhelmingly on potatoes. Potatoes were an easy crop because they required little or no attention (except at harvesttime) and the only equipment needed to grow them was a spade. They were also an excellent source of protein, carbohydrates and mineral salts. An acre and a half of potatoes would feed a family of five or six for a year, as long as there was also a cow to supplement this diet with milk. Supporting the same number of people on grain would have required four times the acreage, as well as expensive equipment and draft animals. Of course, this was a situation fraught with risk. Exclusive reliance on one crop meant that if it were to fail, mass starvation was inevitable — which is exactly what happened.

In 1842 a potato blight was reported along the Atlantic coast of America and another occurred there three years later. It was probably during this second outbreak that a diseased potato crossed the ocean to Europe. By autumn 1845 the disease was active in Ireland, with horrifying consequences. Potatoes that looked healthy when dug up became a mass of stinking corruption within a couple of days.

The spores from the fungus attacking the plants spread epidemically. Within a few months of the first major outbreak in November, Ireland's roads were lined with corpses. Things became so desperate that the living ate the dead. And where famine did not cause starvation, it brought weakness that left the sufferer easy prey to dysentery, cholera, typhus and blindness.

Death and devastation had not been seen on such a scale since the Black Death had ravaged Europe five hundred years earlier. A million people died and another million emigrated. The population of Ireland was halved at a stroke. The disaster forced the British government belatedly to the conclusion that without urgent foreign aid, Ireland was doomed. England was incapable of supplying the amount of food needed, so laws written in the seventeenth century to ban foreign shipping from carrying goods to Britain[3] were repealed in order to permit American warships (with their guns removed) to deliver corn and other food supplies to Irish ports.

The legislation governing which ships could carry goods to England had originally been aimed at protecting English shipowners by giving them a monopoly on transportation. This had been a common practice in Europe during the early years of colonialization because it kept profitable colonial trade in the hands of the mother country. But in the early part of the nineteenth century, with the American colonies gone, Britain's import restrictions had been widened to include the transport of corn, to protect the country's agricultural community from increasingly cheap American crops.

The problem was that Britain was rapidly industrializing, and the population was soaring. Protective import tariffs only succeeded in making things dramatically worse. When cheap corn arrived, say, during an emergency or a shortage, a levy was added to the price, to match that of domestic producers. This was good for local landowners (who had considerable political clout), but bad for the customer (who had no political clout).

But by the time of the Irish famine, these obstacles to free trade[+] 4 *229* *192* (which now affected all commodities) were making life difficult for industrialists, forcing them to buy expensive local goods and materials instead of importing more cheaply. This situation had a negative effect on industrial growth and limited the creation of jobs; so the situation was also affected by social pressures. Thanks to industrialization, the cities were packed with country people looking for work in the factories, and rioting when they found none.

Inevitably, a free-trade movement sprang up; it became known as the Anti–Corn Law League. One of its founders was a Manchester calico manufacturer called Richard Cobden, who put forward a powerfully simple argument. Free trade in corn would reduce the price of bread, which would create employment by lowering wages. Because cheaper wages would make manufacturers more competitive on the international market, they could expand production and hire more workers. Cobden and his colleagues mobilized opinion all over the country, and in 1841 the league decided to contest parliamentary elections on a free-trade platform. Thousands of circulars and pamphlets were distributed all over the country in order to rally support.

The tactic worked, and by 1844 the league's central fund stood at the massive sum of £100,000. New offices were opened in London so as to benefit from direct access to the influential London press. This was the year before the famine in Ireland.

Then the famine struck, and with it came the repeal of the Corn

Laws. But the legislative changes that saved the Irish had only come about because the league had made an incontrovertible case for reform. More than anything, the league had succeeded in generating overwhelming support all over Britain. And in one of those twists for which history is famous, the league had been so successful at canvassing opinion only because it was able to use a totally new way of publicizing the shortcomings of the Corn Laws. This publicity included a number of exposés about payoffs that government ministers had been taking in return for granting individual exemptions to the legislation. The league had spread the word so easily because five years earlier a reform-minded individual named Rowland Hill had, at his own expense, uncovered yet another case of parliamentary corruption.

5 161 *128* Hill had started as a teacher (in Joseph Priestley's[5] Sunday
5 282 *245* school) with vaguely revolutionary tendencies and set up a new
5 306 *278* kind of academy in Birmingham. The Hazelwood School curriculum was modern; and the school was centrally heated, was illuminated with gaslight and included a science laboratory. It made such a splash that many of the leading social reformers of the day
6 189 *153* (Jeremy Bentham,[6] Robert Owen[7] and Thomas Malthus among
7 257 *216* others) came to see it.

Later on these visitors would turn out to be useful allies in Hill's fight for reform, after he uncovered the great parliamentary franking scandal because he had become interested in finding new ways to deliver the mail.

For several years, at his own expense, Hill investigated the postal system. In the course of these inquiries, he found that government officials and members of Parliament were using their free franking privileges to send private packages through the mail. Some of the less-conventional contents included a pair of hounds, a cow, some sides of bacon, two maidservants and a piano. These and other fraudulent postal practices were costing the British Treasury the princely sum of a million pounds a year.

Hill discovered that inefficiency was also wasting time and money on a major scale, hampering business and industry expansion. For instance, postage was charged by the number of sheets of paper in a letter, rather than by its weight, so people wrote all over a single sheet: in the margins, on the back or wherever there was an inch of space. Delivery charges also depended on the distance between sender and recipient. At the same time as they were also delivering the mail, postmen were required to collect payment from senders at their homes or places of business. Consequently, it could take a postman ninety minutes to deliver sixty-seven letters. Hill calculated that a postman could deliver 570 letters in the same ninety minutes if the collection of charges were removed from his duties, and would be less likely to be mugged for his collection money.

Thanks to problems like the ones Hill revealed, many letters were finding their way to destinations by unofficial means. On the first steamship delivery between England and New York,[8] for instance, it was discovered that while five letters were traveling in the official bag, ten thousand letters were being carried in illegal bags on the same boat. Hill suggested streamlining the system. A uniform postal charge would eliminate no fewer than forty different rates. Above all, if delivery were prepaid, it would speed up the service, as long as every house and business had a letter box into which the "free receipt" letters could be dropped. The marquis of Londonderry objected to this idea, on behalf of people who didn't want holes cut in their expensive front doors.

In 1837, when they first became public, Hill's ideas generated parliamentary committees, countrywide meetings, public petitions and articles in *The Times* of London. In August 1839 a bill was passed in Parliament to introduce penny postage, and on January 10, 1840, the first 112,000 letters carrying the new one-penny postage stamp were delivered. That was the day Richard Cobden of the Anti–Corn Law League knew he had found the way to get the league's message out to the public, and he was reported as saying, "There go the Corn Laws." He was right about how postal reform would help his cause. In one week, three and one-half tons of league pamphlets were delivered in Manchester alone. The free-trade reforms were a runaway success. Within two years, the number of letters in the mail had risen to two hundred million, and the idea of a fixed postal rate had spread all over the world.

Between 1840 and 1855 seven billion stamps were printed in Britain. The idea of an adhesive square of paper had originally been only one of a number of approaches to the problem of prepayment. Hill himself had favored prestamped sheets of paper. A public competition to come up with suggestions produced twenty-five

8 237 *199*

hundred entries, including everything from complex embossed-paper designs carrying pictures of Britannia, to the winning idea, the now-familiar "penny black" stamp, printed with a portrait of Queen Victoria. Today these first stamps are collector's items worth thousands of dollars.

The man who won the contract to print the stamps was Jacob Perkins, an American from Newburyport, Massachusetts. Perkins had been in England since 1819, hoping to get a contract to produce banknotes with his new printing process. At a time when forgery[9] was all too common, the main attraction of Perkins's technique was that it enabled an engraver to design notes so complex that they were virtually impossible to copy. In spite of support from provincial English banks, Perkins failed to convince the Bank of England. This was the reason why, in 1839, when Hill was looking for somebody to make his new stamp, Perkins was available.

The American's new printing process was called siderography because the first stage of the work involved engraving a design on a tempered steel plate. The design was then transferred to a soft steel roller by moving the roller back and forth on the plate. Once the design was well impressed, the roller was heat-hardened and used to impress the design onto copperplates, from which the paper was then printed. This technique, known as the American system, printed the penny black and, when the Bank of England finally changed its mind about Perkins's system, gave banknotes the complicated look they have today. Alas, the process never made enough money for Perkins, so he went into the business of designing maritime steam engines for colonial shipping.

By a curious trick of fortune, the colonies had indirectly caused Perkins's printing technique to come into being. In the mid–seventeenth century, when the British had taken woolens to the tropical Molucca Islands to trade for spices, they found — not surprisingly — that the locals preferred lightweight painted cottons from India, so a triangular trade developed. English wool was traded in India (controlled at the time by England) for cotton, which was then traded in the Moluccas for spices to take back to England. A few painted cottons also made their way back to England, the fashion caught on and soon designs were being sent out for Indian artisans to copy.

After a while the fashion for cotton became a craze that began to make a dent in the profits of local British woolen manufacturers. In an attempt to protect domestic industries, pressure was brought on the government to prevent traders from importing foreign cotton. There followed a period of scarcity that, naturally enough, sent demand to new heights. By the mid–eighteenth century the sign of good taste among the well-off (and soon the less well off) was to have matching, multicolored cotton chintz curtains and upholstery. It was in order to tap into this new market that Francis Nixon, an Irishman working in Drumcona, outside Dublin, Ireland, found a way to mass-produce the now-forbidden luxury Indian chintz designs.

The problem with chintz (the Hindu word for "brightly colored") was that the designs were complicated and expensive to make, because each layer of the flower-and-birds-and-clouds-and-country-scene patterns had to be carefully (and separately) printed onto the cloth by a different, single-color wood block, carrying a single part of the design.

In 1760 Nixon came up with a way to print chintz using copperplate. This was the first simple version of the technique that would later be improved by Perkins for use in printing the penny black. Nixon was able to use copperplate printing for patterns as complex as those fashionable in chintz because he had a new thickening agent that would give the dye-paste on the rollers the right consistency, so that it would adhere first to the roller and then to the cloth fibers.

The new color-thickening ingredient was a natural gum obtained from African trees. When dissolved in water, it formed a clear adhesive liquid that held the dye on the rollers. The gum also prevented the moisture in the dye from spreading out in the fabric and creating blurred designs. The most common of the new thickeners was made from Senegal gum, which, by the late seventeenth century, was coming into Europe from Africa in great quantities, thanks to the disastrous state of the French economy and the actions of the man who saved the country from total ruin.

His name was Jean-Baptiste Colbert, the son of a master draper from Rheims (although he claimed to be a descendant of an ancient royal Scottish family). Not excessively bright, he was described as of a "naturally sour countenance, with hollow eyes, thick black eyebrows, an austere appearance, and savage and negative manners. . . ." He also drank to excess. In 1648 he secured his financial position by marrying an heiress for whose father he had conveniently arranged tax exemption. After a meteoric rise in the service of Cardinal Mazarin, Colbert was running the Royal Council (and, effectively, the country) by the time he was forty-two.

Colbert decided to restore the tottering fortunes of France. This task was made more difficult by his boss, one of the most extravagant monarchs who ever lived: Louis XIV,[10] the Sun King, builder of Versailles and the Louvre, and the man who came close to bankrupting the country. In the long run, in spite of the odds, Colbert was to succeed, though at the cost of his reputation. When he died, some versifier (perhaps with a tax problem) wrote:

10 264 *223*
10 274 *235*

> *Colbert is dead, and that lets you know*
> *That France is reduced to the lowest state,*
> *For if anything were still left for him to take,*
> *This thief would not have died.*

Colbert began by setting in motion one of the most thorough renovations of a nation's economy ever seen. One of his tactics for boosting French trade was to offer tax breaks and market monopolies to anybody who could find export-import opportunities abroad. The Dutch, Portuguese, Spanish and English had built up their fleets and had been making money hand over fist in foreign trade for some time,[11] so Colbert decided France should join the club. When the New Senegal Company was set up by a group of investors, he granted it a monopoly on the slave trade. At the time it was believed that the Senegal River connected to the Nile (so it was an easy trade route to Egypt), that there were mountains of gold to be found in the African interior and that Senegal would be an easy place to grow sugarcane, cotton, silk and indigo. In the event, little of this proved true; but among other things that French traders did find there was the Senegal gum which they brought back for Francis Nixon and others to use.

11 157 *126*
11 243 *205*

Meanwhile back in France, Colbert's changes were galvanizing the country. The biggest stumbling block to France's becoming a great sea power was the threadbare French navy. It had been allowed to go to rack and ruin. When Colbert took over, all the naval arsenals were empty and the country possessed only three seaworthy warships (out of a total of twenty-two, all of which were obsolete). What few professional sailors were left all worked on contract abroad, leaving only galley slaves to man the fleet. Colbert set up naval technology centers, commissioned the construction of shipyards to build a hundred new ships, rebuilt harbors, restocked arsenals, established naval colleges to train officers, began regular recruitment and opened schools of hydrography. Ten years after Colbert came to power, the navy budget had grown forty-fivefold.

His policy for domestic trade was no less radical. He rooted out fraud in tax-collecting, and simplified tariffs. He standardized tolls on all roads, imposed heavy import duties to protect French industries, introduced a national road-building program, standardized weights and measures, built fortresses all round the country, reduced the government debt by one-quarter, offered incentives to encourage new industries, forced trades to organize in guilds, established state monopolies (like those of the Gobelins tapestry factory[12] and the tobacco industry) and issued strict rules and regulations to ensure fair practices in every aspect of trade and commerce. By the time Colbert was finished, France was well on the road to economic recovery.

12 123 93

What is perhaps the greatest monument to Colbert's reforms can still be seen and enjoyed today. It is the beautiful Canal du Midi, which runs 161 miles from Toulouse on the Atlantic to Sete, near Marseilles, on the Mediterranean. The Canal du Midi had been part of Colbert's scheme (which also included dredging many rivers to make them navigable) to improve the country's communications, at a time when roads were so bad that water was the only feasible way to move goods around the country. The idea of a canal was originally suggested to Colbert by its designer, Pierre-Paul Riquet, who pointed out that it would offer foreign ships going to and from the Mediterranean the means to avoid the long and dangerous voyage round Spain. In this way, the income from revenues of passage (so far enjoyed only by the king of Spain) could go into French coffers.

In 1666 approval was given for the work to begin. It was a massive and unprecedented undertaking, and it set the style for every European canal that followed. The first-ever artificial reservoir was built to provide water for those sections of the canal too high (the terrain rose to six hundred feet at one point) to be serviced by rivers. Three large aqueducts carried the canal over rivers and ravines, and the first canal tunnel in Europe ran over five hundred feet. The canal had 101 locks, including an extraordinary, eight-step staircase lock just outside the town of Béziers. The Canal du Midi was the wonder of Europe. When it was finally ready for traffic in May 1681, the work had taken eight years and employed twelve thousand men.

One of the engineers working on the canal was another man who left his very visible mark on France, Sébastien Le Prestre de Vauban. At the time, this brilliant career military officer was already famous for the fortresses he had built all round France (which still stand) as part of Colbert's national defense upgrade. But apart from canal building and war, Vauban also tried his hand at gunpowder manufacture, architecture, mining, bridge and road construction, hydrography and surveying. He wrote articles on such diverse subjects as forestry, pig breeding, taxation, colonial policy, religious tolerance, privateering and beekeeping. His wry sense of humor can be seen from time to time. In retirement in 1705, he wrote an article entitled "Various Thoughts of a Man without Much to Do." Vauban also calculated that the 100,000-strong French population of Canada would, by A.D. 2000, rise to fifty-one million. Of Canada he wrote: "Let no man say this is a poor country where one can do no good. . . . There is everything."

Vauban designed and built one of the Canal du Midi's three aqueducts and, when the canal was complete, surveyed it for the king. He also (was there no end to the man's talents?) invented the modern socket bayonet and introduced a new kind of siegecraft. This involved digging a trench parallel to the fortified enemy wall, equipping the trench with mortars to give covering fire, digging another shorter but wider trench perpendicular to the first, installing firepower there, then digging another parallel trench and so on, until the last trench was close enough to the walls for infantry to mount an assault because they would now be exposed only briefly to the defenders' fire on their final dash.

Vauban's new trench siege techniques were to achieve their most spectacular success over a hundred years later, and several thousand miles from France, at Yorktown. This siege ended the American Revolution on October 19, 1781, with the surrender of the 13 230 *194* British to a combined French and American[13] army. Between October 6 and 19 the attackers built two parallels (fifteen hundred men digging, protected by three thousand musketeers) into which were placed over a hundred pieces of artillery. Yorktown was reduced to rubble. The defeated British marched out, fittingly, to the tune of "The World Turned Upside Down," played by the American victors.

But it was not only for the English that the world had been re-arranged. Perhaps those who suffered most as a result of the American victory were the 100,000-odd Americans who had supported the English in the conflict. These so-called loyalists comprised pro-English colonists (like Thomas Hutchinson, governor of Massachusetts) as well as disaffected black slaves, ethnic minorities who had felt threatened by Puritan culture, Anglicans discriminated against by Congregationalist policies, and Southerners. The total number of loyalists added up to about only 20 percent of the population. It was a misplaced British assessment of the size of loyalist support in America that led to so many military failures during the war.

After the defeat, loyalists were considered traitors and met punishment at the hands of the victors, which varied from tarring-and-feathering to expropriation of all their possessions, and even execution. One judge in Virginia was so quick to exact summary retribution that his name gave a new verb to the language: *lynch.* Because of this kind of treatment, after the war over one hundred thousand loyalists left America for Canada and England. Thirty thousand settled in Nova Scotia, joining other refugees crossing the Atlantic, who were fleeing from the Highland clearances in Scotland, where lands were being emptied (by the simple but effective means of eviction and slaughter) to make way for sheep.

Life in Canada was far from easy for the new arrivals, especially for blacks. Unemployment was high, and many — fearing they would be sent back to America and slavery — soon emigrated to Sierra Leone in Africa. Conditions were so grim that Nova Scotia was known as "Nova Scarcity." Epidemics and starvation were common, and punishment for breaking the law was extremely harsh. Whites were punished by fines, blacks by corporal punishment. A woman could receive seventy-eight lashes and a month's hard labor for theft of a pair of shoes. Two counts of petty theft brought two hundred lashes. On one occasion a thief was hanged for stealing old clothing worth perhaps $100 in today's money, and on another the hanging was for stealing a bag of potatoes.

Into this northern demi-paradise came a loyalist called Abraham Kunders, who had operated a small shipping fleet in Philadelphia (and lost it, with everything else he owned, before he fled). In spite of Nova Scotia's reputation, it was better than nothing; so loyalists and destitute Scots were continuing to arrive in great numbers. Somebody had to provide for their transportation, so Kunders found a couple of partners, went into shipbuilding and then started buying steamships — first, in order to ply the American coastline with loyalist passengers and, later, to bring Scots across the Atlantic. But by 1840 the flood of immigrants was beginning to diminish.

14 189 *153*Then Kunders heard about the new penny post[14] starting up in the United Kingdom and realized there was going to be a need for regular transatlantic mail steamers. Within weeks he was in London with proposals that got him the contract. The first Kunders post boat sailed from London to Halifax and then (no fool, Kunders) on down the coast to New York. The new shipping line became known as the British and North American Steam Packet Company. By mid-century the Kunders family had a fleet of thirty vessels and

15 237 *199*
15 299 *269*the number grew, as the new era of the luxury liner[15] dawned. Well into the twentieth century the Kunders line dominated Atlantic passenger routes with big ships like the *Lusitania* and the *Mauretania*, winning blue ribands for speed and setting new standards in luxury and comfort.

In 1969 came the maiden voyage of the line's greatest-ever flagship, the *Queen Elizabeth 2*. By now, of course, the spelling mistake (made by some immigrant official, either in the United States or Canada) had become the new version of the original loyalist shipbuilder's name. The *QE 2* was now a "Cunard" liner. And like all other luxury cruise ships, she had a hairdressing salon where passengers could get a permanent wave.

The engines that power the great cruise liners of the modern world all began with the efforts of a Scotsman who was trying to drain mines. . . .

2 Revolutions

HISTORICAL myths die hard, don't they? In spite of the facts, they persist. Like the one which starts this story: James Watt sitting in his mother's kitchen, watching the kettle boil and dreaming up his great steam engine that would power the Industrial Revolution and change the world.

In fact, the idea came to him in 1765, in the repair shop at the University of Glasgow, for the very prosaic reason that an already-existing model steam engine (used for demonstrations by the Natural Philosophy Department's laboratory) had broken down. Watt fixed it with a minor modification, for which he got all the credit as inventor of the steam engine, totally eclipsing the reputation of Thomas Newcomen, a hardware salesman from Dartmouth and the engine's original designer.

The reason Watt then became famous so quickly was that Britain was going through an economic boom, which generated a growing need for raw materials, especially minerals. Miners were digging ever deeper and, as they did so, getting their feet increasingly wet. Mines were flooding, and Newcomen's engine (in fact, a pump) wouldn't drain them fast enough — until Watt improved the engine, after which everybody wanted one. Watt's career as a manufacturer of steam-driven mine-draining pumps was assured. This suited him fine, because that's all he wanted to be. Neither he nor anybody else had given much thought to other uses for the pump engine. Using it to drive factory machinery was out of the question because the gearing mechanism that would turn an up-

and-down pump motion to a round-and-round driving motion —
the "sun and planet" gear — would not be invented for another sixteen years, by one of Watt's employees, William Murdock[16] (who
got the job because he came to the interview wearing a wooden hat
he had made).

16 59 *45*
16 103 *72*

This gearing system worked simply enough. A fixed gearwheel
was attached to a connecting rod hanging from the steam engine
beam. As the beam end went up and down with the action of the
pump piston attached to its other end, the wheel teeth meshed with
those on another gearwheel, which was attached to the end of a
driveshaft that was free to rotate. As the connecting rod went up
and down, its fixed gearwheel rotated around the drivewheel, the
action looking like a planet circling the Sun.

With this rotary motion, Watt's steam engine could turn wheels
carrying belts and so drive machines in cotton mills, corn mills,
grinding mills, rolling mills, potteries, sawmills, iron foundries
(where the steam engine worked the bellows in blast furnaces),
breweries, starch makers, bleach makers, oil mills and cloth mills.
As for kicking off the Industrial Revolution, by 1795 Watt had introduced all the basic industrial work practices at his new engine-
making factory in Birmingham.[17] The plant was laid out so as to
maximize the flow of production for standardized engines. Jobs
were broken down into specified operations, with appropriate specialization of labor.[18] Watt could then pay piece rates because he
was able to establish how long it should take to make a standardized component.

17 136 *105*
17 221 *184*

18 78 *52*

So, in spite of his inventing the steam engine being a myth, Watt
was indeed the originator of many of the manufacturing techniques
that made possible the Industrial Revolution. But there was another, less well known invention by Watt that was to have as great
an effect, in its way, as his steam engine. One day, in the twentieth
century, it would trigger a revolution as fundamental as that of
the industrial era because Watt's other invention would bring into
common use a kind of soot. In the modern world this material
would open the way to the investigation of the processes of life itself
and trigger a second, biological revolution.

This second revolution will bring radical change to life in the
twenty-first century, and the trail leading to it begins with Watt's
steam engine business and the troublesome fact that he was too

successful. It was while Watt was in Redruth, Cornwall (where many mines were interested in his steam pump because their galleries ran out under the sea, and flooding was particularly common) that he found himself overwhelmed with paperwork for a "multiplicity of orders." His greatest problem was, as he wrote to a friend, "excessive difficulty in finding intelligent managing clerks." In 1780 he came up with a way to solve the problem: an alternative way to make copies of technical drawings, invoices, letters and all the documents that needed duplicating. (He'd already tried and failed with a two-nibbed pen.) The patent for his idea referred to "A New Method of Copying Letters and Other Writings Expeditiously." Watt had invented the copier.

Documents were written or drawn on damp paper with a special ink that included gum arabic, which stayed moist for twenty-four hours, during which copies could be made by pressing another smooth white sheet against the original and transferring the ink marks to the new sheet. Initially, the copier was not a success. Banks were opposed because they thought it would encourage forgery.[19] Countinghouses argued that it would be inconvenient when they were rushed, or "working by candlelight." But by the end of the first year, Watt had sold two hundred examples and had made a great impression with a demonstration at the houses of Parliament, causing such a stir that members had to be reminded they were in session. By 1785 the copier was in common use.

    19 222 *185*

Then in 1823 Cyrus P. Dalkin of Concord, Massachusetts, improved on the technique by using two different materials whose effect on history was to be startling. By rolling a mixture of carbon black and hot paraffin wax onto the back of a sheet of paper, Dalkin invented carbon copies. The development lay relatively unnoticed until the 1868 balloon[20] ascent by Lebbeus H. Rogers, the twenty-one-year-old partner in a biscuit-and-greengrocery firm. His aerial event was being covered by the Associated Press, and in the local newspaper office after the flight, Rogers was interviewed by a reporter who happened to be using Dalkin's carbon paper. Impressed by what he saw, Rogers quit ballooning and biscuits to set up a business producing carbon paper for use in order books, receipt books, invoices, etc. In 1873 he conducted a demonstration for the Remington typewriter company,[21] and the new carbon paper became an instant success.

    20 69 *47*
    20 81 *53*
    20 135 *104*

    21 145 *113*

The paraffin wax Dalkin used, and which was therefore half-responsible (together with carbon black) for changing the world of business, had originally been produced from oil shale rocks. After the discovery of petroleum[22] in Pennsylvania, in 1857, paraffin oil was produced by distillation and was used primarily as an illuminant to make up for the dwindling supply of sperm-whale oil in a rapidly growing lamp market. Chilled-down paraffin solidified into paraffin wax. Apart from its use in lighting, the wax was also used to preserve the crumbling Cleopatra's Needle obelisk in New York's Central Park.

A more everyday use came with a new and exciting way to make fire. For centuries travelers had either carried glowing embers with them or found an already-made fire from which to take a light. But as transportation improved and people traveled farther and faster, these means became impractical. So by the mid–nineteenth century the new phosphorous match had become popular. By far the most successful type of match was the one invented by two Swedish brothers called Lundstrom. Their "safety" match was tipped with red phosphorus, instead of the previously common white version of the mineral, for the very good reason that white phosphorus tends to ignite spontaneously (and also poisoned the matchmakers). In order that their match would burn easily after the initial phosphorous flare, the Lundstroms injected a small amount of paraffin wax into the wooden splint, just below the match head.

Phosphorus had one other, very odd side effect. It gave the British the reputation of grave robbing in order to help solve the problem of feeding a rapidly rising urban population. Thanks to James Watt's steam power and industrialization, the English manufacturing towns were expanding at a breakneck pace. During the nineteenth century population in the cities rose from one-third to four-fifths of the national total. The census of 1851 already showed that for the first time, anywhere in the world, there were more people in towns than in the countryside. One typical cotton town, Oldham in Lancashire, had 12,000 inhabitants in 1801, but by 1901 the number had risen to 147,000. In the same period the national population tripled.

While some of the reason for the population surge was a declining death rate, due to better hygiene[23] and a general improvement in health, most of the increase was related to improvements in diet and plentiful food supplies. This became possible because of phosphorus and the work of a German chemist called Justus von Liebig,[24] whose trick was to burn vegetation to discover its chemical constituents. Liebig thought that plants derived their nutrition from the soil and the air. Using his own money, he set up the world's first real chemical research laboratory at the University of Giessen in Germany. It proved so popular that his students came from all over the world. In his lab, Liebig discovered a law of crop growing that was to have an astonishing effect. The "law of the minimum" states that crop yield is determined by whichever one of a crop's natural nutritional elements is lowest in quantity.

But Liebig's key discovery was that phosphoric acid was necessary for all plants. The easiest way to produce phosphoric acid was to treat ground-up bones with sulfuric acid. The English led the way in this type of production and by 1870 were producing forty thousand tons a year. This output was what led Liebig to accuse them of grave robbing to feed their crowded city populations:

> England is robbing all other countries of the condition of their fertility. Already in her eagerness, she has turned up the battlefields of Leipzig, of Waterloo and of the Crimea; already from the catacombs of Sicily she has carried away the skeletons of many successive generations. Annually she removes from the shores of other countries to her own, the manurial equivalent of three millions and a half of men, whom she takes from us the means of supporting, and squanders down her sewers to the sea. Like a vampire, she hangs around the neck of Europe — nay, of the entire world — and sucks the heart blood from nations....

If indeed this grave robbing really happened at all, it could well have been triggered by Liebig's *Organic Chemistry and Its Applications to Agriculture and Physiology*. The book became an overnight international bestseller, running to seventeen editions in eight languages, and turned agriculture into a science. In the book he showed how ground-up mineral phosphates, treated with sulfuric acid, would produce better fertilizers that would be more easily absorbed by plants. Everywhere the search for phosphates intensified, and in America phosphate fertilizer production went into high gear after the discovery of enormous deposits in South Carolina, Georgia and Florida. Most of the output from these sources went to tobacco growers.

Because of Liebig's work, all over Europe and America in the second half of the nineteenth century, crop yields provided the food so urgently needed for the growing industrial millions. All that remained was to find the means to distribute the food. Thanks again to James Watt, the way was already at hand in the form of the steam locomotive. Although the first steam-driven train (the *Rocket*) had been developed by George Stephenson[25] in 1829 for service on the Manchester-to-Liverpool line, passenger trains had initially met with considerable opposition. Investors considered that the trains offered little hope of profit. Besides, it was said that at 40 mph, the passengers would asphyxiate.

This minor consideration did not, however, prevent the almost incredible expansion of railroads[26] in the United States. By 1838 every eastern state but Vermont had them. By 1850 the network had spread to Kentucky and Ohio. Just after the Civil War, in which railroads played a key role,[27] there were 35,000 miles of track; by 1890 the figure had risen to 164,000.[28] Nothing like it had ever been seen before. From 1869, when the transcontinental line was completed, most railroad company names included the word "western."

Although railroad tracks were used to open up the country and to establish new centers of population, the most spectacular developments came in freight transportation. Mile-long trains rumbled through the night, their whistles echoing mournfully across the land, bringing America's seemingly inexhaustible natural resources and harvests to the manufacturing and population centers of the East. The various railroad companies cooperated to set up more than forty of these fast freight through-lines so that deliveries could go to their destinations nonstop. As a consequence, freight rates fell, and use of the freight services rose from ten billion ton-miles in 1865 to seventy-two billion in 1890. By 1876 over four-fifths of all grain shipments went by rail. Special stockcars were developed for the transportation of live animals. The first refrigeration cars were carrying fresh strawberries east from Illinois as early

25 100 69

26 56 43
26 180 140

27 297 267

28 57 44

as the mid-1870s, and New Yorkers began to see fresh milk again for the first time in decades.

Above all, perhaps, the new railroads (particularly in Europe) enabled people to move around as never before. People began to marry outside their own towns and villages, churning up the gene pool as they did so. The massive increase in the production of coal — to make iron to build locomotives, and then in turn to be used as fuel for the engines — also provided the raw material for the production of coal gas. The gas was a by-product of coal-coking,[29] a technique first made commercially viable by James Watt's assistant, William Murdock (who had invented the "sun and planet" gearing system that allowed Watt's steam pump to drive rotary motion). The new gaslight stimulated more leisure-time reading in general and triggered the birth of the evening class (and unintentionally, perhaps, was the genesis of the educated, professional woman).

So now the economies of the West had well-fed, educated industrial and office workers, as well as efficient sources of raw materials supplying production lines that made goods to be sold by rail-traveling salesmen. In America the only thing that stood in the way of the nation's emergence as the first superpower was the country's lack of an effective way to draw all this together with a communications network. The railroads were to play a central role in solving this problem, though in an unforeseeable and indirect way. In 1851 the problem of running trains in different directions on the same single track (the cause of some spectacular head-on crashes) had been dealt with by the use of the telegraph[30] to organize which train would wait and which would pass. It was only a matter of time until the transmission of Morse code would give way to that of speech, with the telephone, whose development took a major step forward thanks to Thomas Edison,[31] who had spent his early career as a telegraph operator on the railroads.

When the phone went into general use, its chief drawback was that you could hardly hear what callers were shouting into the transmitter at their end of the line. Then Edison thought of using carbon black (the sooty material used earlier by Cyrus P. Dalkin to make his copy paper). There was nothing particularly new about carbon black. The black particles were the finest known particulate and had been used by the ancient Egyptians (and in India and

29 60 45
29 103 72

30 114 85
30 235 197
30 275 237

31 41 32
31 55 43
31 104 72

China) as a black pigment, collected by scraping the residue formed by oil-lamp smoke, to make ink and eye makeup. By the nineteenth century the smoke was being produced first by coal gas and then 32 65 45 from burning coal tar oils, including creosote.[32]

32 140 108 Basically, the telephone worked when a voice vibrated a metal diaphragm in the transmitter mouthpiece. The vibrating diaphragm caused the current in an electromagnet to fluctuate. At the other end of the line, the varying current caused another electromagnet to vibrate, generating a changing magnetic field. This in turn made the metal diaphragm in the receiver fluctuate, 33 51 42 reproducing the original sound.[33] Edison and his backer, Western Union, were looking for a way to raise the sound levels produced by this system when, in 1877, somebody suggested that carbon black was supposed to be sensitive to an electric charge. When subjected to pressure, its electrical resistance changed. So Edison 34 54 43 tried it, first of all separating Bell's[34] receiver from the transmitter (placed in the same box, they caused interference) and then putting a small button of compressed carbon black between the vibrating diaphragm in the transmitter and its electromagnet. The first demonstration to the directors of Western Union caused a sensation. The carbon button worked so well that it was still in telephones fifty years later.

By 1880 the telephone was also bringing dramatic change to the shape of the city, by helping cause suburbs to come into existence. Horse trams had been available for years to take people out from the city center, but there was little incentive to move (especially for business owners) without an effective means of communicating with the downtown headquarters and factories. The telephone provided that means. Industrialization had also caused a boom in land values, substantially raising the cost of living in big houses in the city center. In any case, the newly affluent middle class wanted to get away from the workers, now crowded in tenements around the factories.

Increases in land values also triggered the building of skyscrapers, now that architects and construction bosses could talk by phone to their foremen up and down the building, instead of having to use whistles or messengers. Soon small retailers, feeling the pinch of rising downtown costs, began to move their businesses to the sub-35 115 86 urbs, using the phone to place orders with city center wholesalers.[35]

Thanks to all these changes, by the end of the nineteenth century there was a large and growing suburban market for a more individual form of transportation. Henry Ford[36] answered the call with his Model T. The new cars were soon running on more durable tires, thanks again to carbon black. In 1904 it was found that the mechanical strength and endurance of rubber[37] was powerfully increased by adding carbon black, because it greatly reduced the speed at which the rubber oxidized.

36 166 *130*

37 67 *46*

37 144 *108*

And then came one of those strange twists of history, bringing together the phosphates that had helped to feed the city populations, and the electricity that made possible the telephone which was now giving daily lives new shape. This time, the major change would come because of a scientist who couldn't find a job.

For some time it had been known that passing an electric charge through a piece of metal in a vacuum tube causes it to give off mysterious streams of particles[38] called (after the electrified metal) cathode rays. These rays can be focused through an aperture, into a pencil-thin beam, and then magnetic fields can be used to direct the path of the beam. It was also known that if the rays are directed at a glass plate covered with a phosphorous material, the screen glows where the rays strike.

38 52 *42*

38 239 *203*

At the time scientists were principally interested in this phenomenon because they hoped it might reveal something about the behavior of electricity in a near vacuum. Nobody cared much what the cathode rays might be encouraged to do, so there were no plans to use the rays for any practical purpose. The end of the nineteenth century was the era of the amazing X-ray[39] discovery, and everybody was keen to find other rays in the vacuum that might do similarly miraculous things.

39 116 *88*

39 226 *187*

At this point comes the twist in the tale. Ferdinand Braun, a German physicist, came to the reluctant conclusion that his field (radiation in vacuum tubes) was, to say the least, oversubscribed. Everything that could be done to tubes, electric currents, cathodes and screens had been done. So in 1896 Braun decided to look at the only thing left unresearched: the cathode rays themselves. Some years earlier Heinrich Hertz had shown that an electric current is made up of consecutive and repeating positive and negative cycles, and that the current can be defined by the frequency with which these cycles happen every second. However, nobody had ever actually seen this cycle occurring. Braun realized that cathode rays would make this sight possible and that such a development would allow engineers to monitor electric current in power generation, where it is essential to be sure that the power supply is at a constant, unvarying frequency. So far, there had been no way to do this.

Braun built a vacuum tube with a neck which opened out to a phosphorescent screen. When he set small electromagnets round the neck, he was able to use their fields to move the particle beam around. The field created by the electromagnets could also be made to affect the beam in reaction to the positive-negative changes in the current. In this way, Braun was able to get the beam to cause a spot to move across the phosphorescent screen while it moved up and down in response to the changes in the current, showing the current as a glowing sine wave. Braun's invention — which became known as an oscilloscope — could be used to reveal the characteristics of any current. It was a highly precise analytical tool, and the

40 50 *41* forerunner to the modern television[40] tube, whose picture is built
40 280 *241* up by a beam of particles scanning back and forth, in a sequence of lines from top to bottom, on the screen.

But it was Braun's ability to measure electric current that takes the story to the next step and brings carbon black back into the picture, through Edward Acheson, a twenty-eight-year-old Ameri-

41 31 *29* can who was working for Thomas Edison[41] at Menlo Park. In 1880,
41 55 *43* after working for a period in Europe, Acheson returned to the
41 104 *72* States and began installing electric-light machinery. Since opportunities in the prime electricity-generating industry were pretty well exhausted, Acheson identified a niche market, in the generator-manufacturing business, for industrial abrasives.

His first idea was to make artificial diamonds for abrading tools; so he began to experiment by mixing clay and powdered coke, and fusing the mix at extremely high temperatures in an electric furnace. The result was a compound, silicon carbide, which Acheson named Carborundum and which turned out to be second in hardness only to diamonds. The abrasive characteristics of his new material won him a contract with Westinghouse, the firm supplying

42 73 *50* electric lights for the World's Columbian Exposition of 1893[42] in
42 109 *73* Chicago.

It was when Acheson accidentally overheated his mixture one

day (to a temperature of 7,500°F) that he found that the silicon in the Carborundum had vaporized, leaving him with almost pure graphite. Graphite is a rare form of carbon black, in those days imported from Ceylon, and is extraordinarily resistant to wear and tear, or to extremes of temperature. Acheson promptly found patentable uses for it in electrodes, dynamo brushes and batteries. However, a few decades later Hitler's rocket engineers would find another, more deadly use for graphite.

In October 1942 came the first launch of the Nazi terror weapon, the V-2, whose full name in German meant "Vengeance Weapon 2."[43] The rocket, launched from the pad at Peenemünde on the Baltic Sea, was over forty-seven feet long and nearly five feet across. Lifting off with a thrust of twenty-eight tons, its propellant burn lasted just over a minute and its speed at burnout exceeded 3,600 mph, at a height of 300,000 feet. The V-2 had an initial range of two hundred miles, and from 1944 to the end of the war, over a thousand were launched against England. Hitler's dream was to develop an upgraded version that would reach New York; for this reason above all, graphite was essential — because the burn time for such a rocket would be lengthy. Graphite was the only material that could be used on the aerodynamic control-vanes mounted in the rocket exhaust, because it could take extremely high temperatures over long periods without deforming.

This chapter began with Watt's first revolution. Graphite is now to be key to another. Back in 1895 the excitement over X-rays that had affected Braun's decision to analyze cathode rays also got people interested in what X-rays actually were. Wilhelm Röntgen, their discoverer, thought they might be extremely high-frequency light waves. Unfortunately, the only way to prove this was to see whether X-rays could be made to create interference patterns in the same way that light did. Interference occurs when light bounces simultaneously off a series of surfaces. As the reflected light waves spread out, in a process called diffraction, they interfere with one another, building up or canceling one another and producing characteristic light-and-dark interference patterns. The question is, what would be small enough to act as a series of separate targets off which to bounce the extremely small-wavelength X-rays?

Not many years earlier René Haüy, a French geologist, had

noticed that rock crystals tended to break into regular shapes. When the pieces were further broken, they continued to make smaller and smaller regular shapes. Haüy theorized that in order to do this, crystals had to have regular atomic structures, called lattices. In 1912 it occurred to the German physicist Max von Laue that if crystals did indeed have regular atomic arrangements, the atomic lattice should be able to act as the infinitesimally small, regularly spaced targets off which X-ray waves could be bounced, so as to create interference patterns. Laue's theory presupposed that the X-rays hitting the crystal would cause their electrons to give off "secondary" X-rays which would then interfere with one another (if they were, as Röntgen had believed, light waves). The best crystal for this purpose turned out to be graphite, because its electrons are less tightly bound to the atoms and thus more likely to react to the energy from the incoming X-rays.

The very first experiment showed Laue that he was right. The scattered secondary X-rays fanned out around the central beam, exposing photographic paper[++] when they hit it. Gradually, all round the axis of the main X-ray beam there appeared the familiar light-and-dark interference pattern caused by diffraction. So X-rays *were* a form of light. But there was something immensely more exciting. It became clear from the experiments that diffraction patterns vary depending on the atomic structure of different crystals. So for the first time it was possible to identify a solid material by nondestructive means. X-ray crystallography had been invented.

It was through the use of this technique that in 1952 Francis Crick and James Watson were able to confirm the three-dimensional structure of a molecule of protein. They saw that it took the form of a double helix, which agreed with what they had already deduced chemically. Their X-ray diffraction pattern confirmed the existence of the DNA molecule.

Because of the discovery of DNA, science is already well on the way to the Biological Revolution: developing forms of gene therapy to cure or prevent disorders, and manipulating genes to produce hybrid organisms like tomatoes that have more taste, strawberries that are not damaged by frost or perhaps new forms of animal life. Above all, work is proceeding on deciphering the human genome, the DNA library all humans contain, that makes them who they

44 191 *155*

are: sick or well, black or white, perhaps even intelligent or stupid. The world was not ready for the far-reaching social effects of the Industrial Revolution, triggered in the first place by Watt's steam pump. Is it ready for the Biological Revolution, triggered in the second place by his copier?

The first X-ray diffraction pattern of DNA was visible because of the way X-rays, bouncing off atoms, were made visible in a photograph. . . .

3 Photo Finish

OFTEN on the vast web of innovation and change, things are interdependently linked. One thing is there because of the presence of another. In a modern racing-car event such as that at Le Mans, the moment when the winning car takes the checkered flag is recorded for posterity by the world's press and camera crews. In this case, because of the way the web works, the car which is photographed as it wins the race owes its victory to photography.

The idea of loading film into a camera, snapping the picture and then sending the film to a store to be processed was the brainchild of an American from Rochester, New York, called George Eastman. One day in 1870, at the bank where he had worked since leaving school at the age of fourteen, he didn't get the promotion he was expecting. So he left and used his savings to set himself up as a "Maker and Dealer in Photographic Supplies." At this time, picture taking was a messy, cumbersome and expensive business, involving glass-plate negatives, buckets of chemicals and monster wooden cameras. When Eastman had finished his experiments with the process, his slogan promised, "You press the button. We do the rest."

In 1884, during the early stages of his work, Eastman put gelatin silver bromide emulsion on a roll of paper. This was not a new idea, as Leon Warneke had done the same thing in 1875. The process involved loading the roll of paper into the camera, taking an exposure, developing it, sticking it facedown onto glass and finally stripping off the paper. This also was not a new idea, and it didn't work this time around, either. The paper tore, the emulsion blistered and the pictures were fuzzy. But the basic approach was right.

Then Eastman found the answer. (Some say the deed was done by Hannibal Goodwin, an obscure Episcopalian minister in Newark, New Jersey, with whose family Eastman's lawyers eventually settled for five million dollars.) The Goodwin-Eastman invention was a strip of flexible, transparent, nonflammable material which was the perfect inert base for the photographic chemicals.[45] By 1895 Eastman had produced it in rolls which went into a camera small enough to hold in one's hand — called the Kodak. The new Kodak put photography into the pocket of the amateur-in-the-street and took the company Eastman and his partner founded from a two-person operation to one that employs thirteen thousand people in factories all over the world.

45   108   73

That the wonder material both Goodwin and Eastman had used was available to be made into strips and to revolutionize photography was all thanks to the Great Disappearing African Elephant Scare, reported in the *New York Times* in 1867. Great white hunters all over Africa were said to be shooting the elephants at an alarming rate. A certain Major Rogers of Ceylon was said to have bagged no fewer than two thousand in a lifetime of hunting. The reason for all the mayhem was that the market for ivory in Europe and America had tripled over the previous thirty years. By 1864 England alone imported a million pounds of it annually. Since each elephant tusk averaged sixty pounds in weight, English imports accounted for 8,333⅓ dead animals a year. The ivory so obligingly provided was used mainly for ornaments and for the most popular indoor sport of the day: billiards. The best billiard ball came from the center-line of the best elephant tusk (only one in fifty was good enough). Without a crack or blemish, this ivory was then seasoned for two years. So the news out of the African bush was not good for playboys of the Western world.

With typical American enterprise, in 1869 the billiard ball manufacturers Phelan and Collander offered $10,000 in prize money for an ivory substitute. This offer attracted the attention of a couple of printers in Albany, New York, called John and Isaiah Hyatt. They came up with a substitute that was indistinguishable from the original and that revolutionized not only the world of billiards but also the market for bric-a-brac. The new material softened at temperatures close to 100°C (212°F), so it was easy to shape. It was tough, uniform and resilient, with high tensile strength. It was also resistant to water, oils and dilute acids. At first it replaced natural materials like rubber, gutta-percha, bone, ivory, shell and horn. It could also be made to look like ivory or coral, amber, mother-of-pearl, onyx and marble.

Its uses multiplied to an incredible extent. Just a few of the things for which it was employed were dolls, vases, handles, combs, buttons, musical instruments, canes, knife handles, electric-wire coating, toys, sporting goods, thimbles, fountain pens, stays, ice pitchers, piano keys, checkers, dominoes, dice boxes, cheap jewelry, saltcellars, soap dishes, key rings, thermometers, tape measures, brush handles, mirror backs, pincushions, shaving brushes, button-hooks,[46] glove stretchers, hairpin stands, stick and umbrella handles, shoehorns and trays, as well as boxes for powder (tooth, nail, talcum and glove) and pomade, Vaseline, jewels, salves and cold cream. The new substance also breathed life into a small town in the Jura in France, Oyonnax, which became the European comb-and-sunglasses center of the nineteenth century.

The Hyatt brothers' invention changed the workingman's wardrobe, too. The new material was shaped into collars and cuffs that would never fray or wear out and that always looked freshly ironed. These became de rigueur in banks and telegraph offices. One other outlet for the new product gives a clue as to one of its antecedents. The Hyatts opened a company called Albany Dental, to make substitute ivory false teeth. Part of their sales pitch claimed that as camphor was involved in making the product, the teeth smelled "clean." However, on occasion, as reported by the *New York Times* of 1875, the teeth also exploded.

The Hyatts' invention was called (by Isaiah) celluloid. It had come into existence in 1833 when a French chemist named Henri Braconnot was playing around with nitric acid and potatoes (an activity known at the time as vegetable chemistry). In 1838 a second Frenchman, Théophile Jules Pelouze, substituted paper for the potatoes. Finally, in 1846 Christian Schönbein, professor of chemistry at the University of Basel, perfected the process by replacing the paper and potatoes with cotton wool and adding sulfuric acid, to produce a new secret weapon (but not secret for long): exploding cotton wool.

There were several things that could be done with exploding

cotton wool. One, tried in Boston, Massachusetts, in 1847, was to mix it with ether and turn it into an antiseptic to be placed on open wounds. Another was to use it for waterproofing bearskin hats. Yet another use (the one tried by the Hyatts) was to mix it with camphor, heat it, squeeze it and turn it into celluloid. The most impressive use, however, was the version seized upon eagerly by every military man who could get his hands on it.

Guncotton (as it was called) was a wonder-weapon that had several extremely valuable features. It was three times as powerful as gunpowder and was also smokeless and flashless, so the enemy couldn't see it being fired. The last two attributes alone guaranteed its success among artillerymen. Other minor pluses were that guncotton didn't suffer from the damp, wasn't affected by heat and didn't dirty the gun barrel, all of which were problems normally associated with the use of gunpowder.

Guncotton was first used in 1864, and by 1880 its highly successful reception encouraged the growing use of new brass cartridges. The only real problem with guncotton was that it was *too* explosive. From time to time it would blow up unexpectedly in the factories where it was being made, on one occasion destroying the entire English town of Faversham. Then a rich, pyromaniac Swede called Alfred Nobel[47] (who made much of his fortune from developing the Russian oil fields in Baku) mixed guncotton with ether and alcohol, turning it into nitrocellulose. He mixed that with nitroglycerine, and then mixed everything with sawdust. In 1868 the mix was named dynamite.[48] Shells (and military missiles of all kinds) were soon filling the air more than ever before.

47 138 *107*
47 148 *115*

48 139 *107*
48 149 *116*

This explosive extravagance would one day lead to Concorde airplanes and atomic bombs, because a fellow in Vienna was curious about what happens when a bullet whistles past somebody's ear. It was a well-known fact that when this happens in battle, the survivor is shaken by two bangs. One is caused by the gun. Nobody knew where the other bang came from. Ernst Mach (the Viennese investigator) decided to look for the cause of the second bang. He was a psychologist deeply interested in perception and had read reports of a Hungarian high-school teacher called Antolik who'd done experiments with electric sparks and soot. The sparks were blowing the soot off glass plates, even though they shouldn't have been able to do so. Mach fired a bullet so that it cut two very fine wires stretched across a sooty glass tube. As the bullet cut the wires, it triggered two powerful electric sparks and the shutter of a high-speed camera. The photograph revealed the first spark illuminating the soot disturbed by the V-shaped bow wave of the bullet, and the second spark lighting up the turbulent air the bullet left behind.

Mach worked out that the bow wave happens faster than the speed of sound. (He called this speed "1," hence the modern use of the term "Mach 1" to describe the speed of sound.) The bow wave is a shock wave, and Mach was able to photograph it because of the way the light coming through the glass tube is bent by the turbulent air to form what looks like streaks. This technique is called schlieren photography (the word means "streaks" in German), and it has been used in all aerodynamic experiments since Mach. It was this newly discovered shock-wave effect (now known as the Mach effect) that would one day help physicists to decide the most effective position from which to detonate the Hiroshima bomb.

Mach was interested in seeing if he could make the shock wave visible, because he believed that only phenomena which can be observed, or that are tangible and quantifiable, have any reality. "If you can't sense it," he said, "forget it." So Mach also held that there are no universal absolutes, only local phenomena which people observe subjectively. In this way, his other experiments on perception anticipated astronaut training, because he whirled on a huge rotating mechanism volunteers (wearing paper bags over their heads) strapped into a chair on the end of a twelve-foot arm. Once they were up to speed, the subjects lost the sense of being in a spin, proving Mach's point that perception is entirely subjective. Everything, for Mach, is relative to the observer.

These thoughts were music to the ears of another interested party, Albert Einstein (who was also closely linked to events at Hiroshima).[49] Einstein always attributed his ideas on relativity to Mach's influence, referring to Mach's work as intellectual "mother's milk" for most physicists of the period.

It is also Einstein we have to thank for the fact that, besides the theory of relativity, there is Hollywood and the movie industry. This connection came about because at the beginning of the twentieth century nobody knew what light was. Sometimes light acts as if it were a wave, rippling out in concentric circles from the source, exhibiting a wavelength and a frequency. This behavior can be verified by creating interference patterns. On the other hand, some-

49   227   *188*

times light acts as if it were made up of particles. This was first noticed in 1873, when a telegraph operator at the transatlantic terminal on the Irish Atlantic coast saw that his equipment was giving off electric current according to the amount of sunlight coming through the window. The more sun, the more current. In the evening no current was present at all. It turned out that the light was hitting selenium-metal resistors, so the selenium was obviously giving off electricity in response to light.

There are certain events that form major crossroads for the pathways that crisscross the web of change, and this is one of them. Many journeys on the web pass through such crossroads. James Watt is one; so is infinitesimal calculus; so are the Romantic movement, Newton, printing and coal tar. Similarly, the discovery of selenium represents a major nexus because its discovery was so fundamental that it triggered many subsequent innovations and changes.

For example, by 1884 a German called Paul Nipkow was using selenium on an image-scanning disc, trying to convert pictures to signals which could be sent down a wire. He failed, although what he'd done was establish the technique that would eventually be used for all early television,[50] before the development of electronic scanning. Meanwhile, it was clear that the electric current which selenium was emitting happened not in ways that would relate to waves of light but in discrete bursts of electric charge. These were made of electrons whose numbers increased with the intensity of the light. But in experiments, the charge didn't change when the frequency of the light was changed, as might have been expected if the charge were related in any way to waves of light. Only the speed of the electron discharge increased with an increase in wavelength frequency. Einstein resolved the mystery. Light behaves like waves *and also* like particles (which he called photons). It does so according to the way in which it is measured. The observer can count the particles or calculate the frequency of the wave, but never both at the same time.

This earth-shattering scientific breakthrough was of supreme irrelevance to people in the film industry. For them, the way selenium produces electricity from light meant only one thing: talking pictures. If selenium could be made to give off electrons in response to a fluctuating light produced by a varying electric charge, in turn

50 40 *32*

50 280 *241*

produced by a membrane vibrating in reaction to the way a sound went up and down, then the fluctuating light could produce a dark-and-light print on a moving strip of exposed film. When this film was processed, and light shone through the light-and-dark print, a fluctuating amount of light would get through to hit another selenium cell. When this happened, the cell would emit changing numbers of electrons, which made a varying electric charge, which vibrated a membrane and reproduced the original noise that had made the membrane vibrate in the first place[51] when the sound was being recorded onto the film. That, according to Professor Tykociner of the University of Illinois (who worked on the problem from 1900 to 1918), was how sound-on-film should work. Unfortunately for Tykociner, the sound was too weak to hear in a cinema, so he gave up the research and lost his chance for a place in Hollywood's Hall of Fame.

The problem was solved by a fellow with the Hollywood-sounding name of Lee De Forest, who invented a device without which the world might still be an electromagnetically silent place. It was already known that in a glass lightbulb containing a vacuum and an incandescent filament, a stream of particles[52] passes from the hot filament to the cold metal baseplate of the bulb. De Forest thought this stream could be used like a booster. The charge coming out of the filament is negative, so it heads toward the positive baseplate. But if a small metal grid were placed between the filament and the baseplate, and then the grid were made electrically negative, it would stop the stream of particles by repelling the negative charges (since negative repels negative). But when even the smallest *positive* charge is placed on the grid, the particle stream coming from the filament is powerfully attracted toward the grid, surges through and hits the metal baseplate going at a much higher speed. If the small positive charge onto the grid has been generated by a weak signal (like that given off by selenium), the signal will get a massive boost from the electron stream, and the effect of the boost can be tapped off as a stronger signal when the stream arrives at the metal baseplate. So the "audion" (De Forest's name for his device) boosted the small signals generated by the system for creating sound on film. Now cinema audiences could hear the sound.

And so could anybody with any signal receiver, because De Forest had also solved the major problem of how to make audible the extremely weak signals broadcast by the newly invented radio:[53] by

51 33 30

52 38 31
52 239 203

53 236 198

linking a number of his small audion tubes, weak signals could be amplified millions of times, increasing the signal-reception range several hundredfold. De Forest proved this in 1910 when he transmitted a live Enrico Caruso concert. But De Forest had done much more than provide a new medium for the world of news and entertainment. His amplification technique made possible long-distance communication of every kind. Before the audion, telephone[54] calls couldn't go much farther than two hundred miles. Then, in 1914, thanks to a series of De Forest amplifiers set at intervals to boost the signal along the line, the first New York–San Francisco telephone link was established. Also for the first time, six simultaneous conversations could be carried by the same telephone wire, with each conversation being transmitted at a different frequency. Amplified radio signals would eventually make possible television and satellite communication.

54 34   30
54 276 238

De Forest had the idea for an audion because of an accident that had occurred some years earlier in West Orange, New Jersey, in the laboratory of history's most famous (and most self-publicized) inventor, Thomas Alva Edison.[55] In 1883 he was working on his new, incandescent electric lightbulb when he happened to notice it was getting smutty from carbon deposits building up around the base of the bulb where he had sealed it with a small metal plate. Without realizing that the phenomenon was being produced by a stream of cathode particles, the inventor modestly named it (and even patented it) the Edison Effect. Some years later De Forest would make good use of the effect in his audion, although he always denied Edison the credit.

55 31   29
55 41   32
55 104 72

Edison had been given his start on the road to fame and illumination during his early years as a telegraph operator, with time on his inventive hands. He worked for the railroads, where he'd been constantly threatened with the sack for experimenting with current, circuits, magnets and similar electrical matters. It was the unprecedented expansion of the American railroads[56] that would offer Edison many opportunities, just as it did for many other entrepreneurs and inventors.

56 26   28

As the tracks pushed westward, they opened up new markets everywhere. Whole wooden cities would spring up magically overnight at the mere mention of an approaching railroad construction gang. Entire forests disappeared with the same alacrity, for the same reason. By the mid–nineteenth century, the American forest was the first example in U.S. history of a resource exploited beyond sustainability. If the railroads were to continue expanding at the same accelerating rate, within decades there would be no more American forests. This early environmental problem would turn out to have unexpected consequences.

The railroads used wood (with uncontrolled abandon) for bridges, wagons, carriages, crew accommodations, locomotive fuel (three thousand cords a month), telegraph poles and, above all, the ties that held the rails in place. In 1850 America had 9,000 miles of track. By Edison's time, in 1890, railroad track totaled 164,000 57 28 *28* miles.[57] The gangs were laying the rails at more than ten miles a day. Ten miles of track needed twenty thousand wooden ties; approximately two thousand trees were processed for each day's work. In 1856 the 1,700-foot-long bridge linking Rock Island, Illinois, and Davenport, Iowa, required a million feet of timber. Small wonder that the state of Michigan (whose white pine was a favorite for ties) was logged to exhaustion by the late nineteenth century. The great primeval forests went down before the axe and were then hauled away by dozens of logging railroads to the crews building the line to the Pacific, now that gold had been discovered in Califor- 58 1 *10* nia.[58] In 1850, near Saginaw, Michigan, six lumber mills had cut three million feet of timber a year. By the end of the century, eighty Saginaw mills were cutting thirty million feet annually. By 1856 Illinois was being called "a treeless prairie," and there was no cordwood to be found within thirty miles of Chicago. The lumber trade created many millionaires. One of them, Ezra Cornell, gave up selling plows to go into business providing poles for the Western Union Telegraph Company. In time he also became one of their shareholders and made enough money on the deal to found a university in Ithaca, New York (to which he had earlier built the railroad), and name it after himself.

But the real cause of the timber problem was that it was ongoing. Timber in railroad ties and telegraph poles lasted only five to seven years before rotting and having to be replaced. And then, in 1856, the American forests were saved, thanks to a series of events that began with a newfangled kind of illumination making its first railroad appearance on the Galena and Chicago. It was gaslight. Ironically, given this link with steam engines, the first person to make gaslight a viable and economic proposition (and eventually to save the forests from the depredations of the railroads) had been one

of James Watt's colleagues, William Murdock,[59] who in 1792 had
introduced the first commercial use of coal for gas in England. By
1802 he had installed gas burners at Watt's factory outside Manchester.

 Coal gas was made by the simple process of coking coal.[60] This
gave off fumes which, if filtered, would burn with a yellow light
bright enough to see by at night. One could promenade, safe from
muggers, in the newly gaslit London streets of 1812, go to concerts
in the gaslit Brighton Pavilion in 1821 or, by 1829, read at home in
the evening or work through well-lit night shifts in factories. By
this time coal gas was being supplied by two hundred gaslight companies. By mid-century there were also gas manufacturers in every
urban center of America, where there was a superabundance of the
raw material.

 In England, objectors to gaslight argued that it undercut the
whaling[61] industry. Whale oil had previously been the main source
of fuel for lamps, and since gaslight reduced the need for whales,
it also reduced the number of whalers ready-trained and available
to the Royal Navy. At the time England was busy fighting the Napoleonic Wars, and the navy was badly in need of experienced
sailors.

 The discovery of coal gas also generated another eco-crisis besides the one that had already occurred in the North American forests. The main waste product created by coking coal to make gas
was coal tar,[62] a black, foul-smelling sludge that most gas makers
dumped into the nearest river or pond. By the mid–nineteenth century London's River Thames was so polluted by the tar that Parliament couldn't stand the stink and had to close for business. This
spurred serious attempts to deal with the problem. One set of investigators set about finding out what else could be done with the tar,
besides dumping it.

 Typically for the period, it was German chemists who came up
with the answer: distill the tar. The process created a number of
useful by-products, including kerosene for oil lamps, synthetic colors,[63] antiseptics[64] and aspirin. One other by-product was to save
the American forests: a thick, oily, black liquid called creosote.[65]
When creosote was brushed onto wood, the wood would last
thirty-five years in the open air instead of only seven. This trick
proved so popular that the streets of New Orleans were even paved
with creosoted wood blocks.

By an extraordinary stroke of fate, however, it was the same coal tar now saving the railroads that would one day help to kill them off. Early in the nineteenth century a Scotsman, Charles Macintosh, had found that another coal tar by-product, naphtha, would partially dissolve rubber. So he made a fortune by spreading dissolved rubber between sheets of cotton and creating the waterproof rain-coat[66] still known as a mackintosh in Britain today.

66 144 *108*

The naphtha-dissolving technique created many new and different uses for rubber. But the problem with rubber was that it wasn't a very versatile material.[67] Macintosh found, for example, that in very hot weather his raincoats would "sweat," and in freezing conditions they would crack. The solution to this particular problem came, as ever with innovation, by accident. In 1839 a young American working in the Roxbury India Rubber Company in Roxbury, Massachusetts, was experimenting with his raw materials one day when he accidentally let a mixture of rubber and sulfur drop onto a hot stove. The next morning he saw that the rubber had charred, like leather, instead of melting. He correctly inferred that if he could stop the charring at the right point, he'd have rubber that might *behave* like waterproof leather. The sulfur had vulcanized (he coined the word) the rubber in such a way that it would retain its shape and elasticity over a wide range of temperatures. So now rubber could be hard or elastic, as required.

67 37 *31*

By 1844 our young American had patented the process and begun one of the Western world's great industrial manufacturing companies, which he named after himself: Goodyear. But in spite of the success of the company, Goodyear got into debt (at one point he was in the Paris debtor's prison); and in 1860 he would finally die in New York, in traditional inventor fashion: penniless.

But Goodyear's new product made all the difference to bicycle[68] riders and to Macintosh's fashion-conscious rubberwear customers. By this time the most profitable market for rubber was footwear. Thanks to vulcanizing, boots and shoes were now elastic-sided and rubber-soled. In 1857 a rubber adhesive was produced for sticking on rubber soles preformed in steam-heated vulcanizing molds. In response to the growth of a leisured middle class and the general interest in sports and health triggered by the cholera epidemics of the mid–nineteenth century, the British produced a new type of athletic shoe made with a canvas top and rubber sole. Because the thin

68 77 *50*
68 281 *243*

band covering the join between sole and upper, running the length of the shoe, reminded the manufacturers of the line on the side of a ship indicating its load safety limit, the new shoe was called after the inventor of the loading line: plimsoll. Then in 1865 came the cycling shoe, and in 1876 the baseball shoe. In the 1880s new asphalt tracks and courts generated demand for a rubber running shoe. But what turned the rubber industry into really big business was the American Civil War, with its need for millions of rubberized groundsheets, rainproof garments and tents.

One other aspect of the Civil War was also to affect the future of raincoat material. The Union army achieved notable success with a Balloon Corps,[69] which first used tethered balloons for high-altitude spying on enemy activity at the Battle of Fair Oaks, in 1862. Later, messages could be sent back and forth between signalers in balloons tethered high over hilly terrain. It was in order to watch the Balloon Corps in action that an aristocratic German officer was seconded to the U.S. Federal army. He was so impressed with what he saw that he returned to Germany and designed his own airborne warship and named it after himself: Zeppelin. In 1917, during World War I, the first Zeppelin bombing raids on England brought the full horrors of modern mass warfare to major cities for the first time. And the giant airships were able to operate in virtually any weather because their giant gasbags were made of rubberized raincoat material.

The airship also provided support and encouragement for the designers of its means of propulsion, the new internal combustion engine driving the propellers of the first zeppelin. It was with the arrival of the automobile (and the road-building program it generated, which would change America from a land of railroad users to one of car drivers) that the rubber industry would finally be transformed, and the railroads virtually destroyed.

The key to success in the new rubber market would lie in how fast manufacturers could produce the goods. Once again, coal tar did the trick. In 1856 yet another coal tar by-product was isolated by researchers at the British Royal College of Chemistry in London. It was there that a German professor, August Hofmann[70] (who back in Germany had been a pupil of the great organic chemist Liebig),[71] set up a laboratory where his more able assistants did the practical work, investigating his theory that coal tar was made of a single

69	20	*25*
69	81	*53*
69	135	*104*
70	141	*108*
71	24	*27*

chemical base (known as aniline) from which many more products could be derived.

Some of the first of such products were a whole range of new artificial colorants known as aniline dyes, which were then success-
72   142   *108*
fully marketed by Germans like Friedrich Bayer,[72] who went on to found the great German chemical, pharmaceutical and plastics industry. Then, early in the twentieth century, it was discovered that if aniline were added to rubber, it cut the time for vulcanizing automobile tires and inner tubes by over two-thirds. It also enhanced the serviceability of the finished tire, which would now last much longer in the course of normal motoring — long enough to survive the punishing experience of a track event like the twenty-four-hour race at Le Mans (where the winner is photographed riding to victory on tires made possible, in the first place, by photography).

One of the earliest uses of rubber tires changed the leisure habits of the world when they were fitted to a bicycle. . . .

4 Better Than the Real Thing

MODERN living is facilitated by technology that makes everything convenient. Household appliances provide the kind of help that less than a century ago would have needed a dozen servants. Today's homemakers have at their fingertips more power than any Roman emperor. The laptop computer gives a modern individual more data-processing capability than the total available to the Allies in World War II.

However, the price to be paid for all this technology is the frenetic pace of modern life. There is no time for the leisurely business lunch of yesteryear. In a busy working day there is no time to spend hours at a table. Fast food is sold ready-cooked, to be heated in the microwave oven. And the packaging blurb assures consumers that the dish is even healthier for them than the fresh food they no longer have time to prepare. Thanks to the marvels of modern science, the instant meal (with all its artificial flavoring and added nutritional value) is even better than the real thing.

The chain of events described in this chapter — which led to the modern convenience food we now consume in ever-increasing quantities — began, appropriately enough, with an invention that was itself designed to make life more convenient. In 1893 Whitcomb L. Judson of Chicago patented a new gizmo called a clasp locker, also designed for use with money belts and tobacco pouches. The system was an adaptation of the hook-and-eye system previously used to fasten dresses, and Judson merely added a clasp lock at one end for security. The clasp locker caused a sensation at the

73 42 *32*
73 109 *73*
74 26 *28*
74 181 *141*

World's Columbian Exposition of 1893.[73] In 1908 Gideon Sund-back, a Swedish electrical engineer who had earlier worked for the Westinghouse generator works in Pittsburgh,[74] was managing a hook-and-eye-making factory in New York when he filed a patent for the separable fastener. We today call Sundback's invention the zipper. By 1918 the U.S. Navy was using zippers on sailors' uniforms, and the British rag trade was putting them on skirts and dresses. By 1930 the zipper was in universal use.

75 46 *38*

Interestingly, the original reason for Judson's 1893 version was revealed by another of its names: a shoe fastening. The period, known as the Naughty Nineties, was characterized by the way skirt hemlines rose, daringly, to reveal the ankle. Because of this overexposure, high-button boots became fashionable among respectable ladies.[75] Some boots had as many as twenty buttons. Judson's idea was to cut down on the time it took to put these monsters on. The market for boots had been growing since mid-century, as typewriters, telephone exchanges, printing machines, telegraphs and a gen-

76 146 *114*

eral business boom created more jobs for women,[76] put money in everybody's pockets and created new leisure industries.

77 68 *46*
77 281 *243*

One of the most successful of these industries was bicycle[77] manufacture. Cycling became a craze for both sexes — and so did new cycling fashions. A new garment was promptly designed to solve the awkward problem of cycling in a long skirt. The brainchild of Amelia Bloomer of Seneca, New York, *culottes* became all the rage, and soon any woman with pretensions to fashion consciousness was wearing bloomers. However, they revealed even more ankle and required even higher boots. This ever-changing boot-and-shoe market was to generate even more jobs for women.

The man responsible was a bigamist who had eighteen children by five women, three of whom were married to him at the same time. In order to make sure he never made a slip, he called all the daughters Mary. This lovable rogue was a mechanic-turned-actor, whose performances were described as "crude and bombastic." From acting he would finally turn to industry and social climbing.

In 1844, after fifteen years of wandering and womanizing from New York to Rochester, Baltimore and Chicago, Isaac Merrit Singer ended up briefly in Fredericksburg, Ohio, where he found himself with a job carving wooden letters for typesetters. Inventing a machine that would do the same work, he took his prototype to

Pittsburgh. But the machine failed to sell; so he moved on to New York and then finally to Boston, where he settled for a while.

On the floor above the apartment where he lived was a man called Phelps who made sewing machines. Singer became interested in the machines because there appeared to be money in them, and after looking at the various examples available on the market, he came up with two modifications that made him a fortune. He added a foot treadle attached to a crank turning a belted drivewheel and a "foot," which pressed down on the cloth as it was being sewn. These two additions were critical to the efficient function of the machine, but they were not the reason why Singer became rich.

What made all the money was the marketing talent of Singer's partner, Edwin Clark, who probably did much more than Singer to make the sewing machine a major international world-changer, with new sales techniques that the business world has used ever since. It was Clark, for instance, who thought up the idea of the *Singer Gazette*, distributed free to all customers and carrying ads about Singer and his business methods. It was Clark who introduced the idea of buying machines by putting five dollars down and making monthly payments with interest. He also suggested the idea of part exchange on old machines, aimed all Singer advertising at women, employed women demonstrators, persuaded church groups to buy machines at a discount (thus making the machines respectable) and convinced husbands that the machine would give their wives more free time.

Clark's efforts succeeded beyond anyone's wildest dreams. In 1856 the company made 2,564 machines. Four years later it had sold over 100,000. By 1861 Singer was already selling more machines in Europe than in the United States and six years later, the Singer Corporation had become the first multinational, with factories in Britain, France and Argentina. Singer's machine also started mass-produced styles and then the mail-order business that soon sprang up to merchandize them. Visitors to America remarked on the way Singer had democratized fashion. American shop assistants even looked as well dressed as their customers.

The link to the high-button boot was that Singer's machine would be used by bootmakers later in the century to supply customers with the wide variety of footwear styles that appeared with the bicycle craze and other new leisure-time outdoor activities. The sewing machine was sturdy enough to handle leather, so by 1858 it was already being used to sew uppers. Demand generated by the Civil War brought technical modifications that would sew soles, welts, heels and toes. Where skilled hand-lasters could finish only sixty pairs of shoes a day, these new machines made over four hundred. And they would also do ornamental stitching, lace holes and button holes. In spite of all the ballyhoo and sales gimmicks, however, Singer's astonishing success was ultimately due to the fact that he was able to turn out thousands of sewing machines cheaply and quickly (and repair them just as fast), because they were made of identical, interchangeable parts and assembled on a production

78 18 24   line.[78] So if a piece broke, it could easily be replaced.

The sewing machine was not the first piece of equipment to be made in this way. New England clockmakers had been mass-producing parts for wooden clocks since late in the eighteenth century. They worked their pieces on pole-lathes, powered by a length of rope wound round the shaft on which the piece was mounted. One end of the rope was attached to a springy pole and the other to a foot treadle. When the rope was pulled down by pressure on the treadle, the pole bent and the movement of the rope spun the shaft carrying the piece. When the treadle was released, the pole straightened up, pulling the rope and spinning the shaft in the opposite direction. The workers used a cutting tool to work the piece on every other turn.

In mid–nineteenth century New Haven, Connecticut, a clockmaker called Jerome Chauncey began making the first mass-produced metal clock parts, probably using techniques developed by Eli Whitney, who lived in the same town. Back in the 1790s, Whitney, in debt and desperate for money, had been trying to get a contract with the U.S. government (since it was likely to pay its debts) and had talked it into believing he could make muskets. Previously he had invented and manufactured a cotton gin that automatically removed the seed from the plant fibers. This innovation

79 102 72   reduced labor costs substantially and made the price of cotton[79] from America's South so competitive that the giant British textile industry took all of it in preference to Indian cotton.

In 1798 Whitney came across reports of House and Senate debates on the government's intention to procure arms and wrote the secretary of the treasury that since the cotton-gin market was saturated, he now had a workforce (and the know-how) to manufacture between ten thousand and fifteen thousand muskets with "certain machinery moved by water." The truth was that his gin-making factory had burned down (along with twenty gins ready for sale, and all his tools and materials). As for Whitney's gun-making capability, the machinery was not yet built and he had never made a musket in his life. This didn't seem to matter much to the government, which, keen for a quick result, was attracted by Whitney's promise to make muskets with identical, interchangeable parts that could be replaced easily on the battlefield. After a smoke-and-mirrors demonstration of interchangeability to the authorities (all he did was replace a few whole locks with a screwdriver), Whitney got the contract and $134,000 to take back to New Haven and start work.

By 1801 he had delivered the first batch of guns, thanks to Thomas Jefferson's returning from France and sending him a pamphlet telling him how to do it. Jefferson[80] had been in Paris between 1784 and 1789 and had met a man named Honore Le Blanc, who had shown him how to take apart several of his newly invented musket locks and then put the pieces back together again at random.

Jefferson's first job in Paris had been American commissioner with responsibility to negotiate trade treaties between the new republic and the various European powers. After a year abroad he was made minister to France and began to move in elite intellectual circles. Jefferson had always been keen on science in general (especially agricultural science and meteorology), and during his European travels he took notes on innovations he came across: water screws in England, rice production in Italy, canals in France and new construction techniques in Amsterdam. While he was in Paris he also took an interest in ballooning[81] and viticulture (stocking up with the finest Burgundies and muscats). As minister his circle included many of the leading thinkers of the day, like Condorcet (with whom Jefferson had discussions about a future French constitution) and de La Rochefoucauld.

In 1786 he met a man whose work he had known for some years, George-Louis Leclerc, otherwise known as Count Buffon. By this time Buffon was a revered figure in French scientific circles, owing to his years as director of the Jardin du Roi and his monumental, multivolume study of natural history. Only thirty-six of the fifty

80	113	83
80	213	172
80	295	261

81	20	25
81	69	47
81	135	104

volumes were published in Buffon's lifetime, the most famous of which was the fifth, *Natural Epochs*, published in 1778, in which Buffon set out geological history in a series of stages. In it he suggested for the first time that the planets had been created in a collision between the Sun and a comet. He also divided the history of the Earth into seven parts, corresponding to the seven days in Genesis, assuming that each stage was about thirty-five thousand years long.

Buffon was also fond of criticizing America and encouraged other French scientists (who had not visited the United States, either) to do the same. Some of the wilder French publications included allegations by various scientists that climatic conditions in America had produced animals inferior in every way (particularly in size) to those in Europe. It was also claimed that American natives were underdeveloped weaklings who lacked sex drive; that immense areas of America were covered by dense and putrid fogs; that the country was overrun by lizards, serpents and monstrous insects; that frogs in Louisiana weighed thirty-seven pounds; and that syphilis was endemic.

Jefferson's response to this nonsense was to ask a friend, John Sullivan, governor of New Hampshire, to send him the bones and skin of a moose, elk, caribou and large deer. Jefferson then exhibited these at the Cabinet du Roi, as proof that American animals were not undersized. Buffon promised to correct his errors in the next edition of *Natural History*, but he died before this could be done.

Like everybody else in scientific circles, both Jefferson and Buffon were interested in newly discovered animals (like some of the fossils being found in America) because their bones might provide evidence that would complete the gaps in the "Great Chain of Being." This construct had been around since Aristotle had invented it two thousand years earlier. It was based on the belief that all forms of life had been established at the moment of Creation and were arranged in ascending order of quality, proceeding by infinitesimally small stages from simple slime all the way up through humans to angels, who were closest to God. Since the chain had been created in its entirety from the very beginning, any gaps between species that looked too big to be natural had to represent organisms humankind had not yet discovered.

In the previous century the scholars of the English Royal Society had felt it their duty "to follow all the links of this chain, till all their secrets were open to our minds; and their works advanc'd or imitated by our hands. This is truly to command the world; to rank all the varieties and degrees of things so orderly upon one another; that standing on the top of them, we may perfectly behold all that are below, and make them all serviceable to the quiet and peace and plenty of Man's life." Success in identifying and naming every organism on the chain would reveal the totality of God's plan.

What exercised the minds of the seventeenth-century thinkers was how "imperceptible shadings" were involved in the links between one species and another on the chain. How many divisions and subdivisions were there? Were there intermediate stages, halfway between one species and another? This matter of graduation would one day give rise to the concept of evolution, but in the seventeenth century, "gaps" which might appear in the chain were viewed merely as imperfections in human knowledge.

One of the most powerful exponents of this view was the German mathematician and philosopher Gottfried Leibniz,[82] who sought to identify forms of life that might be so small as to escape detection (for example, in the apparent gaps in the chain between some of the species and their marginally different variants). Leibniz suggested the existence of infinitesimally small entities which he named "monads," the smallest particles of existence. Leibniz felt that it was impossible for the senses to grasp the subtly different characteristics of some organisms and their variants, or "where one begins or ends." Leibniz believed there was an "infinity of living things whose small size conceals them from ordinary observation."

Leibniz was interested in minutiae because he had developed a type of math that would help to calculate the infinitesimally small. At the time astronomers were particularly interested in such matters because the recent discovery of gravitational influences on planetary movement made it necessary to work out the accelerating rates of change in the velocity of heavenly bodies in solar orbit. Leibniz's system for accomplishing this became known as "infinitesimal calculus."

Part of Leibniz's reason for believing that there were, indeed, organisms too small to see might have been due to the fact that he had been to the Dutch city of Delft, visiting the grand old man of microscopy, a bureaucrat named Anton van Leeuwenhoek,[83] who had recently made the most amazing discoveries through a new form of magnifying glass that he had invented. Leeuwenhoek had been a draper for some years and earlier, in 1667, while in England

82 231 *194*
82 253 *214*
82 309 *279*

83 254 *214*
83 310 *279*

on business, had seen extraordinary drawings of shot-silk fibers magnified many more times than was possible with his draper's glass. The device that had been used to produce these new views was called a microscope. When Leeuwenhoek developed a more powerful version, he changed the world because he used it to look for life forms too small for the naked eye to see.

Up to this point, microscopists like Nehemiah Grew, Marcello Malpighi and Robert Hooke had examined inanimate objects such as plant sections, corks, dissected animal veins, bronchial tubes, fish scales and so on. But when Leeuwenhoek looked through his microscope and saw tiny objects moving around, he decided that they were alive. This assumption would turn out to be critically important to the development of biology. Using a glass that magnified about two hundred and seventy times, Leeuwenhoek began to see an entirely new world.

In 1676 he caused a sensation at the English Royal Society in London with an eighteen-page letter about the "little animals" he had discovered. He described living protozoa and rupturing cells releasing their protoplasmic contents. In three hundred subsequent letters he included drawings of bacteria, rotifers, spermatozoa, bee and louse stings, liver flukes, frog parasites, plaque, blood corpuscles, embryos and algae and described the copulation procedures of mites. In one letter, in 1677, he announced the discovery of the smallest "animals," thirty million of which, he said, would take up the space of a single grain of sand.

Leeuwenhoek's discoveries astounded everybody, including, in 1690, a young man as different in character and circumstance from the self-made, semi-literate draper in Delft as it was possible to be. His name was Christian Huygens, the well-traveled son of a rich and prominent Dutch family living in The Hague. The Huygens family was decidedly upper-crust. They had entertained Descartes, and the father had an English knighthood. Huygens himself had studied math and law at the University of Leiden, spent time in Paris and London (where he met Newton and other important people) and was now an avid experimenter. After developing skills as a lens grinder (lens grinding was a hi-tech hobby at the time), he built a telescope with his brother and in 1655 discovered Titan and the rings of Saturn.[84] By 1656 he had developed the first pendulum clock and the theory explaining how it worked, and then he

84 132 98
84 252 214

went on to design the balance spring. Huygens also dabbled in the cutting-edge disciplines of the day: math, hydrostatics, astronomy, mechanistic philosophy, ballistics and cosmology.

It was lens grinding that led this urbane sophisticate to the investigation of how lenses work. His research resulted in a paper suggesting that light is a series of shock waves, spreading in wave fronts generated by impacting light particles and making waves among the tiny particles of the "ether," an invisible, intangible, weightless substance permeating everything in the universe.

In 1677 Huygens noticed the strange things a crystal called Iceland spar did to light. It split a light-ray into two parts, one of which was refracted in the way (worked out by Huygens) that glass normally affected light. The other ray obeyed no known law. This, Huygens decided, was because the crystal was made of two different media, one of which changed the direction of the light in an abnormal way. What Huygens couldn't explain was the fact that if the "abnormal" ray then went through a second crystal, the orientation of that crystal would affect the light in a bizarre way. If the second crystal were rotated 360°, the light-ray gradually faded to nothing and then returned to full strength again.

What Huygens didn't know was that the "abnormal" ray was polarized light. Passage through the first crystal made the light waves vibrate in only a single plane. In Edinburgh, Scotland, in 1828, this property of Iceland spar (otherwise known as calcite) was turned to advantage by a Scottish geologist, William Nicol. Previously he had improved on techniques used by a local lapidary called Sanderson, to produce extremely thin sections of crystals and rocks. With these he was able microscopically to study the interior structure of minerals. However, the technique remained unknown because Nicol published only two minor papers (read by almost nobody) on tiny cavities which he had discovered in certain rocks.

In 1828 Nicol split a piece of Iceland spar and then cemented the two parts together again, using a resin called Canadian balsam, obtained from fir trees. Because of the refractive index of the balsam, a polarized ray of light can pass directly through the spar crystal, while the balsam reflects away the ray of ordinary light. If a second, similarly prepared piece of spar is placed next to the first, rotating it dims or brightens the emerging polarized light. This instrument, called the "Nicol prism," reveals that if the polarized light from the first piece of spar passes through certain substances, their structure affects the beam of polarized light by dimming it. The amount of rotation by the second spar crystal in the Nicol prism necessary to return the affected polarized ray to full strength depends on the structure of the substance under examination. So, the amount of rotation identifies its chemical composition. Substances that behave in this way became described as "optically active." In 1840, using this technique, Nicol's prism was used to analyze sugar because the amount of sucrose in the sugar would affect polarized light passing through a sugar solution.

Despite the fact that at this time sugar was a commodity as valuable and expensive as petroleum today (and so you would have expected Nicol's breakthrough to make news), sugar analysis was not how he was to become known. Fame came about because of the efforts of James Neilson, a Scottish furnace manager. In 1824, while running the Glasgow gasworks, he had patented an idea that would change the course of Scottish history. He simply passed through a red-hot pipe the draft of cold air normally used to fan the flames of a blast furnace, in the process heating the air to 300°C (572°F). This caused the coal to burn so efficiently that three times more iron could be made with the same amount of coal. More important, the technique also made it a viable proposition to use the blackband ironstone coal found in abundance in Lanarkshire, but which up to then had been too low-grade to use. By 1835 every ironworks in Scotland was using Neilson's technique, turning Scotland into a major industrial producer. Between 1829 and 1845 the annual production of pig iron rose from 29,000 tons to 475,000 tons, and Scottish iron exports included material for bridges, drinking fountains, sugar-making equipment[85] and even ornamental inkstands. Naturally enough, this surge in industrial production also triggered a coal-mining boom; and that is what finally brought fame to Nicol and his transparent slices.

What happened next involved an amateur botanist from northern England, Henry Witham, who was crazy about the interior of ancient, fossilized vegetables. In 1829, while in Edinburgh delivering a groundbreaking paper entitled "On the Vegetation of the First Period of the Ancient World," Witham met Nicol, and they discussed the latter's thin-slice technique. Witham was excited by the idea of being able to look closely at his beloved ancient plant

85 218 *183*

interiors, so he applied Nicol's technique to the dozens of fossilized samples (of vegetables, fish, teeth and trees) that Scottish coal miners were beginning to dig up. In 1831 Witham wrote up his findings in one of those articles you can't pick up, called "Observations on Some Fossil Vegetables," in which he gave all due credit to Nicol.

The article was read by another obsessive named Henry Sorby (who came from Sheffield in Yorkshire, took his mother everywhere with him and never married). Overwhelmed by the cavities in rocks that Nicol had discovered, Sorby laid the foundations of microscopic petrography and became known as "the man who took the microscope to the mountains." Sorby was interested in the rock structures themselves rather than in any fossil organisms they might have contained, so in 1849 he started to use Nicol's thin-slice technique, together with the Nicol prism, and started discovering things about granite that provided geologists with vital data about the origins of igneous rocks. Sorby had also spent time investigating how sand and silt settled in turbulent water, so he now looked for tiny, telltale ripple-and-eddy marks in stones, to see if they would show him how sedimentary rocks had originally been deposited. On the basis of this work, he was able to tell whether the microcavities had been formed by water or by heat or by pressure. It was this information that would prove so valuable to geologists and others investigating the age of the Earth.

In 1864, when shown a new invention called a spectrometer,[86] Sorby promptly put it together with a microscope and invented the spectrum microscope. His use of the new instrument involved dissolving an unknown substance in water, shining a light through it and then passing the light through a prism. The materials in the mystery substance would absorb the light at certain wavelengths, and as a result, the spectrum visible through the prism would contain black, vertical lines at the point in the spectrum where the color of those wavelengths would normally be seen. These lines would indicate what materials in the substance had absorbed the light.

86 105 *72*
86 198 *157*

Sorby did this kind of spectroscopic analysis on everything, including organic poisons, birds' eggs, plants, algae, human hair, fungi and furnace fumes. It was in 1867 that he turned his attention to autumn leaves, to find out why they had turned yellow-red. He discovered that it was because of a substance which became visible only in autumn, when the green chlorophyll in the leaves disintegrated from lack of sunshine. Sorby called this new, yellow-red material carotene (because it gave carrots their distinctive color). Carotene turned out to be the chemical that gives *all* plants and animals any red or yellow color they might have. Other places in which carotene was found included the feathers of flamingos, lobster shells, apricots, egg yolks and tomatoes. Then, thanks to the work of Franz Boll, a German professor living in Rome in 1876, carotene was also found in the human eye.

Boll was a Berlin physiologist specializing in electric fish when, in 1873, he got an offer he couldn't refuse: to direct the laboratory of comparative anatomy at Rome University. Three years later he was studying frog retinas when he noticed that in strong light the normally red retinal pigment bleached to white. Later that year he went to Berlin and demonstrated this phenomenon to senior scientists like Hermann von Helmholtz[87] and Ernst Pringsheim.[88] After talking to Helmholtz, Boll became interested in the mechanism by which the retinal chemical bleached. The chemical involved would later become known as rhodopsin. Boll's microscopic examination showed that the color change he had observed is due to alterations in the platelets in the outer layer of the retinal rods. As these tissues are always red in dark-adapted animals, but bleached in animals that spend all their time in bright light, Boll reckoned that the red is constantly being destroyed by light and that some kind of "nutrient" restores it each time the eye returns to low light. The "nutrient" would turn out to be a form of carotene, and carotene deficiency would turn out to inhibit the synthesis of retinal purple, causing night blindness. However, this kind of dietary deficiency aspect became known only because of the odd behavior of some Javanese chickens.

The chickens were first noticed in Java by a young army doctor called Christiaan Eijkman, the assistant to a Dutch army commission investigating the causes of beriberi, a debilitating tropical disease affecting Dutch colonial administrators. In 1886 the commission was convinced that beriberi was caused by an infectious agent of some kind. After gathering evidence, the commission returned to Holland, leaving Eijkman in charge of the medical school. It was at this point he noticed something very strange. He saw some chickens in the hospital compound staggering about with

87 233 *196*
88 279 *241*

symptoms remarkably like those of beriberi. Eijkman did nothing about it until one day, for no apparent reason, the birds suddenly lost their symptoms and returned to health. Investigation revealed this event had coincided with the arrival of a new hospital cook, who had refused to feed the birds with leftovers from the staff table because they included expensive polished rice, usually fed to Europeans. Instead, the cook started giving the chickens the unpolished rice eaten by the native Javanese. Once on the unpolished rice diet, the birds became healthy. It was clear that something in the unpolished rice was essential to the birds' health, since without it they had caught beriberi. After analysis back in Holland, Eijkman and his colleagues announced that the essential ingredient was in the rice hulls, which were removed during polishing. But what was the essential ingredient?

The outbreak of World War I brought the answer. In England at the time, 80 percent of the nation's wheat supplies were being imported from America. By the spring of 1917 German submarines were sinking so much supply shipping (to that date, a total of two million tons) that the British national cereal reserves had been reduced to only three or four weeks' worth. Consequently, on January 1, 1918, the government was obliged to introduce food rationing. In preparation, a Royal Society committee had estimated that the average healthy individual needed just over four thousand calories a day to survive. The rationed distribution of meat, butter, margarine and then other foods was related to these figures. But the key question was whether administrators were rationing the right foods to guarantee a healthy diet.

This matter was investigated by Frederick Gowland Hopkins, whose career as a chemist had been highlighted to this point by his work on wing pigmentation in the Brimstone butterfly. Since a key element in this process was the action of uric acid, Hopkins had become an expert in urine chemistry, from which he had developed an interest in the relationship between the constituents of urine and those of diet. From there, it was a short step to the investigation of dietary protein.

In 1911, after a series of experiments on rats, he produced the epic paper "Feeding Experiments Illustrating the Importance of Accessory Food Factors in Normal Dietaries." What Hopkins had discovered (and for which he would receive the Nobel Prize, together with Eijkman) was that even if rats are fed more food than they need, without milk in their diet they die. The "accessory food factor" in the milk turned out to be the protein tryptophan. Not long afterward, Hopkins served on the wartime Royal Society committee on rationing and, as part of his work, looked more generally into accessory food factors and what they might be. In 1919 the committee issued a report. Its title included the new name for the food factors: *The Practical Importance of Vitamins.*

As part of his nutritional research, Hopkins investigated margarine because there were concerns now that it was rationed and in short supply. Hopkins showed that the manufacturing process severely reduced the amount of vitamin A in margarine. Because of this work, by 1926 the first vitamin-enriched margarine was available in shops. It was the first of a series of such additives to be introduced into foods over the following decades, as more and more vitamins were discovered (and named with alphabetic letters because their chemical makeup was not fully understood). Today, virtually every food is vitamin-enriched and, like the first "improved" version of margarine, is usually regarded as being better than the real thing.

Any improvement on the taste of food is always popular. In the Middle Ages, one such improvement changed history. . . .

5 Hot Pickle

THE marketplace has a profound effect on how change comes about. If enough people want hot pickle and will pay any price for it, somebody else will go to extraordinary lengths to find hot pickle for them. In the Middle Ages, for instance, wars were fought over pepper, cinnamon and other such spices that today we throw into the pot without a second thought. Back then they were worth a king's ransom.

It was the laws of inheritance that triggered the medieval European craze for spices, back in the eleventh century. At a time when life was already insecure enough, legislation made sure that estates were passed on intact by limiting the right of inheritance to the eldest son. In those days, land was the prime source of wealth, so there was little for younger sons to do about making their way in life, other than enter the church. So, as is often the case, those who didn't become clerics found themselves with time on their hands, often drifting into bad company. At first, the authorities' answer to the brawling and general hooliganism was to organize jousting sessions. At these tournaments, young landless squires could earn enough to set themselves up, simply by knocking others down. The minimal property requirement for entry to the knightly classes was possession of the kind of stuff easily gotten on a jousting field: lance, sword, saddle, armor and the other accoutrements that went with war. All one had to do to obtain these items was to vanquish (or, depending on the rules of the tournament, kill) somebody else who already possessed them.

But in the eleventh century, the really clever solution to this minor social problem was the Crusades.[89] In one fell swoop, these foreign adventures took younger sons off the street and put them on a boat to the Middle East.

89 300 272

Persuading young lads to go along with this idea was less difficult than it might seem. Taking up the cross and helping to take the Holy Land back into Christian stewardship brought certain rewards, some of them spiritual. Equally attractive, however, was the prospect of loot and pillage. Crusaders were allowed to keep anything they picked up along the way. Easy pickings make for happy looters (which, in the main, characterized Crusaders rather well).

But the way to Jerusalem was strewn with hardship, the worst of which was the matter of getting there at all, on leaky ships and through bandit-infested areas like the Balkans. Many hundreds of young Crusaders never made it past places we now refer to as Croatia and Montenegro. Those who made it in one piece generally took time off (for rest and recuperation) in the greatest red-light district of the medieval world. Constantinople, the glittering supercity center of the Byzantine Empire, must have knocked the arriving Crusaders (from their dung-filled, rat-infested grubby little villages) totally sideways. The place was filled with decadence, intrigue, Eastern promise and, most important of all, the kind of heavenly food nobody had ever tasted before.

It should be recalled that, at the time, European cuisine would have been best described as almost inedible. In winter there was nothing but old salted meat or fish and a few roots to eat. In summer, more often than not, meat and fish were available, but they were usually rotting — since there was no means of keeping them fresh. Food was so dreadful that it was generally referred to only as "that which goes with bread." So when the Crusaders experienced their first taste of Eastern spices, they couldn't get enough.

The spice trade was not new in the Mediterranean. In the first century, out of the eighty-seven different classes of goods imported to imperial Rome from Asia and the east coast of Africa, forty-four were spices. By that time Roman merchant ships were sailing back from India with cinnamon and pepper, officially described as one of the empire's "essential luxuries." A century later caravans were bringing ginger along the newly opened Silk Road running be-

tween Rome and China.[90] Spices were so valuable that in the fifth 90 244 *206*
century A.D., one Roman city under siege by barbarians bought
them off with three thousand pounds of pepper.

After the Dark Ages, when communications were re-established
between Europe and the Middle East, it was spices which stimulated the resumption of international trade; and the commodity was
so valuable that it was as negotiable as silver in the settlement of
debts. Venice was the new entrepôt between East and West, acting
as go-between for the transshipment of pepper, cinnamon, ginger,
cloves and saffron. The returning Crusaders had created an insatiable demand. So when Constantinople finally fell to the Turks in
1453,[91] spice-trade costs went through the roof. Freight charges 91 249 *211*
rocketed, overland travel was too dangerous and, at one point, the
retail price of spices reached ten times the cost at source. Some
other, less expensive way had to be found. This was ultimately the
problem that would kick off Western colonialism.

In the fifteenth century the Portuguese (and then the Spanish,
Dutch, English and French) found new routes to the East, so as to
go and get their own spices instead of paying Turkish profiteers.
The new passage involved sailing round the tip of Africa into the
Indian Ocean, or round the bottom of South America and across
the Pacific. In 1418 Vasco da Gama had reached Calicut, India, and
returned with spices worth sixty times the cost of the trip. Both
the Dutch and English set up large trading companies whose only
purpose was trade with the East.

The profits from these ventures were staggering, rarely lower
than a hundred-percent return and usually more than twice that.
Sir Francis Drake[92] took one ship on an expedition and returned 92 262 *221*
with cargo worth more than Queen Elizabeth's entire revenues for
a year. The opportunity to get at the source of this wealth was too
good to pass up, so by the end of the seventeenth century all the
major Western powers had colonies in India and Southeast Asia.

By that time, however, the taste craze had switched away from
spices. European (and especially English) agriculture had come up
with ways to produce fresh meat year-round, thanks to new kinds
of winter fodder like turnips. Crop rotation was also improving
yields, and there was generally more food available. What was now
delighting European palates was sweet tea.[93] A delectable new 93 2 *11*
luxury called sugar was arriving in increasing quantities from the

94 251 *212* Caribbean and Brazil.[94] Tea and sugar created a market every bit as insatiable as had been the one for spices.

Sugar was so profitable a commodity that the Dutch swapped New York for sugar-growing Surinam, and the French abandoned Canada in return for the canefields of Guadeloupe. China tea had first appeared in Europe in the middle of the seventeenth century and by the eighteenth century, particularly in Holland and England, had become an indispensable necessity. When the English removed almost all tax from tea imports in 1784, the effect was extraordinary. In two years, annual imports rose from £6,000,000 to £20,000,000, much of it going to America.

The only problem was that the Chinese merchants supplying the

95 125 *94* tea insisted on being paid in gold and silver bullion.[95] Since the only English goods to sell successfully on the Chinese market were chiming watches, clocks and music boxes, a serious trade deficit soon developed. In 1793 Lord Macauley went to China as ambassador, carrying samples of other British industrial goods, only to be told by Emperor Ch'ien Lung, "Strange and costly objects do not interest me.... We possess all things. I set no value on strange and ingenious objects, and have no use for your country's manufactures."

But some way to pay for the tea had to be found, because the deficit was hurting the British economy. The situation got steadily worse. Between 1761 and 1800 the British bought goods worth £34,000,000 in Canton (90 percent of it tea), but in return sold only £13,000,000 of British goods. For a while, sales of Indian cotton helped, but then in 1823 China began to produce her own.

Fortunately for the British, who at the end of the eighteenth

96 287 *251* century were busy fighting a war with Napoleon[96] (and were therefore extremely short of bullion with which to buy tea), an alternative was at hand. In India there was one particular product used as an intoxicant by the aristocracy, taken by soldiers before battle to give them courage and consumed by many ordinary people simply to give them energy and endurance in their daily toil. The product

97 207 *166* in question was opium,[97] and the poppies from which it was extracted grew in profusion all over India (at this time controlled by the British).

In China the drug had been known for some time as a medicine but because of its addictive properties, its use was strictly controlled by the Chinese government. Imports were illegal, and the British East India Company (which had a trading monopoly with

China) was warned that it would lose its trading privileges if it brought opium into the country. So the company simply found intermediaries to do the job. Sometimes they labeled their cargo "saltpeter," and sometimes they met offshore with Chinese smugglers who transferred the opium to their junks. Whatever the case, the British could plausibly deny their involvement. Early in the nineteenth century so much Indian opium was entering China that the earlier bullion drain had been reversed.

However, this meant that the opium trade was now an essential element in British state finances. By 1800 London was concerned about French and Dutch interest in the Strait of Malacca and the waters around Java, through which the opium cargoes had to pass en route from India to China. The first English reaction to the danger was to strengthen the British presence in Penang, off the coast of Malaya. This was the event which brought into prominence a young administrative high-flier working for the East India Company.

Stamford Raffles arrived in Penang in 1805 to join the governor's staff. He soon astonished everybody by learning the Malay language, working hard and spending little time in social pursuits. Instead, he got to know local Malays and soon had a reputation among them for fair dealing. He even dressed like a local and ate their food. In 1807 he was made the governor's chief secretary, so his further promotion looked assured. As the result of an impressive report he wrote on Malacca, his next job was as agent to the governor-general in Malaya, where he was given the delicate task of sounding out the Javanese on a possible British takeover of their island from the Dutch. They liked the idea, so in 1811 a British force set out for Java, with Raffles on board the leading frigate.

The invasion went well, and Raffles was soon installed as lieutenant-governor. He promptly began endless tours of the country, partly on official business and partly to indulge in his obsession for collecting natural history. If it grew, flew, swam, ran or swung, Raffles collected it. When he finally returned to London from the East, he had accumulated a vast assortment: two thousand drawings, papers, books and maps, and thousands of specimens both alive and dead, including one of the largest and most foul-smelling flowers in the world, which he had discovered and named *Rafflesia arnoldi*.

A passion for natural history was fashionable at the time,

because in 1802 the subject had been made extremely popular by William Paley, a virtually unknown vicar from the north of England who'd written a bestseller, *Natural Theology*. The intellectual argument of the day was whether the existence of God could be proved by rational argument, and there were hundreds of treatises on the subject. But Paley outdid them all with his "nature-by-design" approach.

His most famous analogy was that of "nature like a watch," made up of many different parts, each of which had a specific purpose and all of which went together to make up a functioning whole. For Paley every part of nature had a purpose. Each part of an animal, for instance, performed a particular function relating to the organism's special needs. Paley gave examples: a woodpecker's beak, the oil gland that birds used to waterproof their plumage, the crane's long legs (since it had no webbed feet and therefore could not swim, God had given it long legs so that it could wade) and the spider's web (it could not fly after its prey, so it trapped them instead). Paley's conclusion was that these function-oriented features argued the case for a designer-God.

98 155 *124* The excitement caused by Paley's book boosted the already-growing interest in taxonomy (John Ray,[98] the originator of plant taxonomy, was one of Paley's heroes) and especially in the eighteenth-century Swedish botanist Carolus Linnaeus's work. He had cataloged and named organisms because he claimed it would reveal God's original plan for Nature. With regard to animals, the interest expressed itself in the new science of zoology and the idea that animals might be collected for study. Raffles had returned briefly to England in 1817 and (no doubt with his collection in mind) had suggested to the eminent naturalist Joseph Banks that there should be a scientific society dedicated to the subject. On Raffles's final return home in 1824, this idea was canvassed by Sir Humphry Davy, another eminent scientist. Thanks to Davy, in 1826 Raffles was named first president of the new "Zoological, or Noah's Ark, Society" of London. Finally, in 1828 the gates to the zoo opened (for members only). Indeed, as one correspondent to a London journal noted, the society took great care "to prevent the contamination of the Zoological Garden by the admission of the poorer classes of society."

Sir Humphry Davy was one of the most extravagantly fêted scientists of the day. When he was twenty-three, his electrical research had already earned him the job of assistant lecturer in chemistry at the London Royal Institution. There he began a series of wildly successful lectures on galvanism,[99] as well as on dyeing and the chemistry of agriculture, which gained him the institution's professorship of chemistry. By the age of twenty-eight he was already a Fellow of the Royal Society. For Davy's work on electrochemistry (he founded the science), Napoleon awarded him a medal, in spite of the fact that France and England were at war at the time. In 1812 he was knighted and became a VIP.

99 184 *145*

99 216 *179*

Davy took up the matter of safety in mines. In 1813 ninety-two deaths had been caused by a terrible methane explosion in a colliery at Gateshead-on-Tyne, in northern England. After a series of experiments to determine the conditions under which methane would explode, Davy designed the miner's safety lamp which today bears his name. He found that if the lamp flame were surrounded by a fine wire gauze, the heat of the flame would not ignite any gas in the surrounding air. For this invention he received a prize of two thousand pounds and the commendation of the Royal Society.

A semi-literate colliery brakeman who had claimed that he, too, had invented a similar lamp at the same time was denied a patent or any recognition, despite the fact that his lamp was already at work in mines and operated on essentially the same principle (his flame was surrounded by a metal shield with holes in it). His supporters were so infuriated that they took up a collection and awarded him a second prize of a thousand pounds. The money not only mollified him but helped to fund work that would make a far greater mark on history than he might ever have made with a miner's lamp — because his next bright idea helped to solve the problem of Britain's massive expenditure on the Napoleonic Wars.

One of the effects of the war was to trigger major inflation and rapidly rising prices. Mine owners in particular were desperate for a cheaper way to transport their coal, now that the price of fodder for draft horses had gone up like a rocket. Ironically "rocket" was the name of the new invention that would move the coal faster and cheaper. With it, the failed lamp maker, whose name was George Stephenson,[100] would change the world. The reason the *Rocket* was so successful with the coal-mine owners was that it used their own product as fuel. So as its use spread (for transporting things other than coal), it also became one of their biggest customers.

100 25 *28*

The *Rocket* was a locomotive, and it was superior to other early

attempts at locomotive design because of Stephenson's revolutionary boiler. The secret of success with a locomotive was to produce
as much high-pressure steam as possible. Stephenson achieved this
by designing a series of twenty-five copper tubes carrying hot water from the boiler through the hot gases escaping up the chimney from the firebox. The tubes provided a much greater heating
surface and consequently generated a great deal of high-pressure
steam with which to drive the engine cylinders, whose pistons
drove cranks to turn the locomotive wheels.

In 1829 at a trial to be held on a section of track at Rainhill,
the Liverpool and Manchester Railway directors offered a five-
hundred-pound prize for the best locomotive. The *Rocket* won
hands-down. Another engine, the *Novelty*, came in last in the competition, and this failure so discouraged its Swedish engineer-
designer that he emigrated to New York. John Ericsson's failure at
Rainhill would turn out to be a decisive factor in the conduct of the
American Civil War.

Apart from locomotives, Ericsson's other talent lay in screws. He
had already built a screw-driven ship, the *Robert F. Stockton*, which
had crossed the Atlantic successfully (although the ship relied on
sail). It was this maritime connection which brought him into contact with the U.S. Department of the Navy. In 1861, at the outbreak
of the Civil War, Ericsson wrote to President Lincoln offering to
help with the building of ironclad ships for the Federal navy. Ever
101 174 *136* since the Crimean War[101] a decade earlier, the vulnerability of
wooden ships to shellfire had become a matter of general military
concern. Ericsson's offer was to build a radically new kind of ship
called the *Monitor*. Lincoln agreed and in October 1861, at the Continental Iron Works in Brooklyn, the first keel was laid. A hundred
days later the ship was complete, and at the end of January 1862
she was launched.

The *Monitor* was unlike any other ship before or since. She dis-
placed 987 tons, carried two 11-inch guns on a cylindrical revolving
turret and made six knots with her steam-powered propeller. She
was also ironclad from stem to stern. But her most unusual feature
was that she operated semi-submerged, with an extremely shallow
draft of only eleven feet. The *Monitor* was ideal for Lincoln's plan
to blockade the South's coastline, because she could maneuver even
in shallow river estuaries, and her very low line made her a difficult

target for onshore batteries. Her revolving turret also made 360° fire possible even while the ship was stationary.

After her first engagement in the Hampton Roads, when she took on the South's battleship *Merrimac*, every navy in Europe wanted one; and Lincoln ordered the construction of another six. A blockade made more effective by the use of the *Monitor* would help prevent the cotton exports which were the major source of Confederate war funding. As part of blockade planning, a number of Southern ports were targeted for occupation. One was Port Royal, on the Sea Islands off the South Carolina coast. Taken by Union forces in 1861, Port Royal became a bunkering and repair base for the Northern navy.

From 1862 Port Royal was the site of one of the most unusual social experiments in U.S. history. Known as the Port Royal Experiment, it gave the island's ten thousand black slaves their freedom and the right to elect their own local government. Sixteen thousand acres of land abandoned by the slaves' previous masters had been confiscated during the war by the Bureau of Refugees, Freedmen and Abandoned Lands, and these were made available freehold for $1.25 an acre at long-term, low-interest rates. Each family's holdings were limited to a maximum of forty acres. Vocational schools, run by graduates from Yale, Harvard and Brown Universities, were set up to teach reading, spelling, writing, geography, sewing and arithmetic.

After the war was over, from 1865 on, the white plantation owners began to trickle back. Intense lobbying on their behalf in Washington (and the perceived failure by the blacks to farm the land profitably) eventually persuaded the authorities to allow "pardoned" white Southerners to repossess most of the land. A few smallholdings remained in black hands, but overall, the experiment was judged a failure. So the Sea Islands went back to producing the Sea Island cotton for which they were already famous. Although the cotton had originated in the West Indies, by the late eighteenth century it was being successfully grown on the South Carolina coast. Its luxury status came from the fact that the creamy white fibers were strong, silky-textured and over 1¾ inches long. In the mid–nineteenth century Sea Island cotton was principally used for underwear and ladies' petticoats (at one time, in the American South, it was de rigueur for women to wear no fewer than sixteen petticoats).

Throughout this period the Southern economy had come to depend almost exclusively on cotton, most of which was shipped to Britain, where the Industrial Revolution *was* cotton manufacture. In 1780 there had been a hundred cotton mills in England. By 1830 there were over a thousand, and the number was rising fast. From about 1820, 80 percent of British cotton imports came from America, where slave labor and the cotton gin[102] had reduced prices to a level with which India could not compete. And although the Civil War temporarily disrupted supplies, by 1880 production was back to normal again, and the mills thundered through day and night, thanks to gaslight.

Since the invention of steam power, all the largest cotton towns had been sited on coalfields, mainly in Lancashire in the English Midlands. Consequently, there was a ready, local supply of the raw material for making coal gas.[103] Gaslight made possible night-shift work and doubled production at a stroke.

It also drastically reduced the risk of fire in mills, because one gas jet would provide the light-level that had previously required twenty-three candles. By the mid–nineteenth century, gaslight had spread far and wide. It was lighting most major city streets in Britain, and reached Toronto in 1850 and Tokyo in 1872. It was this potentially lucrative urban market for cheap illumination that attracted the attention of Thomas Edison,[104] who claimed he would make electricity so cheap that only the rich would dine by candle-light. In 1882 he opened the first electric-light power station, on Pearl Street in New York. The event was a disaster for the gaslight industry, and investment in gaslight shares plummeted. Three years later, an Austrian who had studied under Robert Bunsen[105] (so he knew about gas) found a way to extend the life of gaslight companies by several decades.

Carl Auer von Welsbach was the son of the director of the imperial printing press in Vienna. While at Bunsen's laboratory in Heidelberg, Welsbach became interested in studying "rare earths." It was during this work that he became aware that some rare earths glow extremely brightly when heated. After a number of experiments, he found that if he impregnated a piece of Sea Island cotton webbing with thorium nitrate and the rare earth cerium,[106] and then placed it in a gas flame, the webbing would glow incandescent.

In 1885 Welsbach patented this idea as the gas mantle. It in-

102 79 *52*

103 29 *29*
103 60 *45*

104 31 *29*
104 41 *32*
104 55 *43*

105 86 *59*
105 196 *156*

106 311 *282*

creased the illumination value of gaslight seven times, helped factory owners[107] to keep cotton production high and costs low and 107 290 *255*
kept the gas industry going until just before World War I, when
electricity finally made it too expensive. Gas mantles are still used
today in gas-fueled camping lights. Welsbach spent the rest of his
life investigating and discovering more rare earths, but official recognition (when it came in the form of an Austrian baronetcy) was
for his original work on the mantle. With a perfectly straight face,
Welsbach chose as his aristocratic family motto More Light.

Meanwhile, particularly in the textile-manufacturing industries,
the increasing variety of products was bringing more demand for
the means to control the environment in which the threads were
made, spun and woven. Temperature and humidity regulation was
particularly important. The first major air conditioning system was
installed in American textile factories as early as 1838, with water
spray nozzles operated by a rotary pump humidifying the air and
keeping the threads from breaking. In 1890 air fanned over ice
helped to keep Carnegie Hall cool, and Eastman Kodak[108] film 108 45 *37*
stock safe.

But it was the cotton manufacturers who could really claim to
have invented air conditioning, a term first used by a cotton maker
called Stuart W. Cramer in a paper delivered in 1907 to the American Cotton Manufacturers' Association. The expression referred
not to the air but to the effect of humidification on the condition of
the cotton. The first scientifically designed system for air conditioning (in the modern sense of the term) was outlined in 1902 by
Willis Carrier, an engineer from Buffalo, New York. Carrier circulated cold water through copper coils, blew air over them and controlled room temperature by changing the temperature of the coils
and the speed of the air.

But the real breakthrough in temperature control came with the
development of efficient insulation, and this initially owed its success to the two famous "glass dresses," one of which was designed
for the Spanish princess Eulalia. The other, made for the Broadway
star Georgia Cayvan, was exhibited at the World's Columbian Exposition of 1893[109] in Chicago and is still in the Museum of Arts 109 42 *32*
and Crafts in Toledo, Ohio. The dresses were a great hit with the 109 73 *50*
public, although they had been created more for their publicity
value than with any hope of turning them into ready-to-wear

garments. In appearance they looked like very shiny satin, the glass fibers being interwoven with silk. Besides, they had cost a great deal of money to make. The manufacturer was Edward Drummond Libbey, who owned a glassmaking company in Toledo and employed a young superintendent called Michael Owens. Together with Libbey, Owens went on to set up the Owens Bottle Machine Company in 1903 and later on, one of the biggest glassmaking concerns in the United States: Owens-Illinois.

But what kicked glass manufacture into high gear was demand during World War I for high-quality glass in binoculars, cameras, artillery-fire control instruments and motion-picture projector lenses. Up to then, most American glass of this quality had been imported from Germany; so when war with Germany started, American stocks rapidly ran out. The U.S. National Research Council decided that in the name of national security, at least six different types of glass should be produced by a domestic manufacturer and that military procurement would total at least two thousand pounds of glass a day. Glass manufacture quickly became large-scale and scientific.

It was after the war (when Owens-Illinois and the other American glass giant, Corning, began to cooperate on research) that fiberglass production made major advances, with new glass yarns that could be woven for use in electrical insulation and filter cloths. The same experiments created glass wool, an extremely lightweight insulation blanket. Fiberglass was made by spinning molten glass through tiny holes in a heated platinum box or crucible. As the glass threads emerged, they were twisted together and wound on drums. This product could be woven and then molded into solid shapes. During World War II it was used for nonmagnetic mines, jettisonable fuel tanks and thermal insulation of all kinds.

After World War II the Owens-Corning Company took its new product to the public with a road show featuring alarm clocks ringing unheard inside fiberglass containers, ice cream that remained frozen in its fiberglass insulation while next to it (in an oven) a pie was baked, demonstrations of fiberglass strength with a heavyweight football player swinging from thin bands of fiberglass or an audience member being invited to try (no one succeeded) to break a piece of fiberglass with blows from a sledgehammer.

It was in 1951, while sitting on a park bench in Washington, D.C., that Charles Townes had the idea that would radically change the way glass fiber was used and, incidentally, take up where Welsbach had left off. Townes was an expert on undersea radio-wave transmission and, while sitting on his bench waiting for an appointment with the Office of Naval Research, he was struck by the thought that molecules might be made to vibrate and release microwave radiation. Back in his lab he decided to try to excite ammonia molecules, which vibrate at 240 billion times a second when they are energized by heat or electricity. Townes's idea was to expose these excited molecules to a beam of microwaves also vibrating at the same frequency. This would cause the ammonia molecules to give off microwaves with added energetic input. These microwaves would in turn strike other ammonia molecules and excite them to give off microwaves and so on, in a kind of chain reaction.

By 1953, Townes's experiment had led to a trigger beam of microwaves being produced by a process called microwave amplification by stimulated emission of radiation (MASER). In 1960 Theodore Maiman took the process to the next stage by using the molecules of a ruby. Using a ruby cylinder whose polished, flat, parallel ends were covered with silver, Maiman fed energy into the ruby from a xenon flash lamp. The clear sides of the ruby cylinder let the light in, and because the ruby molecules could be magnetically tuned to vibrate at the frequency of the incoming light, the two interacted to produce the familiar burst of microwave emission, on a single frequency. This emission bounced back and forth between the silvered ends, increasing in intensity and finally producing an incredibly powerful, coherent beam of light.

This intensity-increasing process became known as light amplification by simulated emission of radiation (LASER). More experiments revealed that the best material to use is not ruby, but extremely pure glass doped with neodymium, which is one of the rare earths that Welsbach had discovered and tried in his gas mantle. A neodymium laser could then be used as a source of extremely powerful light to excite other materials, to produce even more powerful beams of coherent light.

In the modern world this extraordinarily new light source is so coherent that during the early Apollo missions it was shown that the light beam spreads only a few feet over the distance from the

Earth to the Moon. Laser light is also so powerful that it will cut steel or the human retina with equal ease and precision. Today, laser light carries digitized signals of all kinds across fiber-optic networks in the world of communications. It is used to measure distances with extreme accuracy. Laser endoscopes light up the interior of the body to aid noninvasive surgery. Laser light shatters kidney stones without damaging other organs. Lasers make holograms, detect the tiniest levels of pollution in gases,[110] scan the interior of the brain, identify flaws in materials, give advance warning of fires and map from space the small shifts in the Earth's surface that herald an earthquake.

But perhaps the most spectacular use of lasers in recent times has been what they have achieved on the battlefield, particularly in the form of the so-called smart bomb. During the Gulf War they were used with devastating effect: bombs were released by an aircraft and then guided to within millimeters of their target, following the hair-thin beam of light, emitted by a laser carried in a different aircraft whose mission was to illuminate the precise location of the target. Ironically, for a story which begins in the Middle East with the medieval spice trade, the onboard switch controlling the release of a smart bomb was called a pickle. And when armed, ready to be used, it was said to be a "hot" pickle.

Smart bombs are among the latest in the military arsenal of quick reaction forces operating in the new post–Cold War environment, where flexibility of response is essential. It was ever thus. . . .

110 197 *157*
110 240 *205*

6 Flexible Response

S UCCESS in war tends to go to the army most able to respond appropriately to any situation, shift ground at will, deliver massive firepower with extreme precision exactly where and when needed, and then withdraw it smoothly and without loss, to be used again elsewhere. Such flexibility of response has always been the key to victory.

One of the earliest and most powerful weapons capable of such flexible application was the bow. Ironically, it was the versatile use of one particular kind of bow which began a sequence of events that would lead to the modern invention that offers the kind of flexible military response without which we may not survive the confusion of the post–Cold War era. The historical weapon is the great medieval longbow.

Nobody knows where the longbow came from, but first references to it point to Wales. Gerald de Barri (otherwise known by his pen name of Gerald of Wales) was one of those medieval priests and administrators who wrote fascinating chronicles of their life and times. In 1188 Gerald accompanied Archbishop Baldwin on his journey through Wales to gather support for the Third Crusade to the Holy Land. Gerald wrote an account of the trip, full of curiosities and local tales. One such story was the report of an event that had apparently occurred six years earlier at the siege of Abergavenny Castle: "Two soldiers ran over a bridge to take refuge in one of the Castle towers. Welsh archers, shooting from behind them, drove their arrows into the oak door of the tower with such force that the arrowheads penetrated the wood of the door, which was nearly a hand thick; and the arrows were preserved in that door as a memento." Gerald actually saw the arrows in question and recounted another story about the same kind of bow. In this one an English soldier had been wounded by an arrow that went through his leg armor, penetrated his thigh and the saddle, and finally killed his horse.

By the fourteenth century the English kings were using archery battalions made up of Welsh bowmen. Their shooting technique and the four-hundred-yard range of the longbow made these men formidable adversaries. They stood sideways to the enemy, so they were hard to hit. A skilled archer could loose up to fifteen arrows a minute; with ranks of archers stepping forward to fire while their colleagues prepared the next shot, a constant storm of arrows could rain on the enemy with devastating effect. It was calculated that during the course of eight hours at the Anglo-French Battle of Crécy, over half a million arrows were shot. The great tactical value of the archer battalion was that bowmen could run to any position, reacting quickly to rapidly changing circumstances in ways that cavalry and pikemen never could, thanks to the light weight and extreme range of their bows.

Longbowmen were medieval superstars; probably the most famous in history was Robin Hood, with his legendary ability to split an arrow from several hundred yards. There are more myths about Robin Hood than about Jesse James. The first reference to the man in Lincoln green was in Langland's poem "Piers Plowman," written around 1377 and referring to folk rhymes about him that were already in circulation.

Robin Hood has been portrayed as many different men. In one version he's a charming rogue, outlawed in Nottingham's Sherwood Forest with his merry band of men, robbing from the rich to give to the poor. This is probably the least credible version of all, as there is no evidence to back it up. Elsewhere, he's a folk-myth hero involved in May Day fertility celebrations. He also figures as a dispossessed nobleman fighting to regain his inheritance; an English guerrilla fighting the Norman invaders; a Crusader in the service of King Richard, fighting for justice after Richard's death and the accession to the throne of his evil brother, King John; and an aristocratic courtier to King Edward II. He also turns up as Robin

Hod, a Yorkshire fugitive from justice living in the woods around Barnsdale.

So far, these various versions boil down to a kind of folk hero, capable of extraordinary deeds of courage, expert beyond all others in the use of his weapon. In this sense, Robin Hood represents a theme that recurs again and again in folk literature since ancient Mesopotamia: the utopian superman who comes to free ordinary people from the humdrum pain of their daily lives. This kind of story is particularly common in societies where the dispensation of justice is arbitrary, and redress of grievances virtually impossible.

The least glamorous version of the Robin Hood legend (and probably the one closest to the truth) comes in a reference recently found in administrative documents relating to activities at the Court of Justices in the village of Eyre in Berkshire. These court records refer to a fugitive called William Robehod who had previously been indicted, in 1261, as one of a criminal gang of three men and two women suspected of robberies.

But whoever Robin Hood really was (and there is little doubt that he did exist at some time in the thirteenth century), what probably occurred was the same thing that happened later in the case of Jesse James. So many people used his name that he turned up in many different places at many different times.

So in all likelihood, Robin is no myth. But Maid Marian undoubtedly *is*. In spite of the tradition that associates her with him, no original medieval version of his adventures makes any mention of her. In fact, the story of their relationship doesn't appear at all until the sixteenth century. The whole romance between Robin and Marian was probably based on a musical written in 1283 in the town of Arras, Flanders. But the work had absolutely nothing to do with England or greenwood forests or outlaws. The author was Adam de la Halle, a married cleric who may have studied polyphony in Paris and who ended up as court musician to Count Robert II of Artois, the province of which Arras was the capital. Chances are that the musical (or, more properly, *pastourelle*) was written to entertain Adam's master's troops when they were all in southern Italy with the count, who was there to lend military support to his cousin, King Charles d'Anjou of Naples.

De la Halle's musical gave the homesick troops what you would expect: a romantic, amusing and slightly naughty view of life back home. It followed the traditional form of the *pastourelle*, a literary form as old as ancient Greece, generally featuring encounters between simple village folk (usually shepherds and shepherdesses) and city slickers who try to cheat or seduce them. In de la Halle's version, the heroine (who just manages to escape seduction at the hands of a knight from the city) is called Marion. The work has the traditional *pastourelle* song-and-dance interludes, interspersed with various episodes of a fairly simple tale: Marion is persuaded by the knight's charms to leave her flock, realizes in the nick of time that his intentions are not honorable and then returns — safe and sound — to the arms of her rustic boyfriend, whose name is Robin.

There was work in Arras for a man like de la Halle because at that time the city was a thriving community of more than twenty thousand inhabitants and already supported two hundred poets and musicians. To a certain extent, the "Robin and Marion" *pastourelle* is a social commentary on the relationship between villagers out in the countryside, who tended the sheep, and the townsfolk (like those in Arras), who profited from the wool. The wealth of Arras, as in much of Flanders, was built on wool. Since the early Middle Ages, the flatlands bordering the North Sea and crisscrossed by 111 203 *163* easily navigable rivers had made Flanders a trading center[111] and a major sheep-rearing area. Textiles dominated medieval European industry at the time, and Flanders dominated European textiles. When de la Halle was writing, Arras and the other great textile cities like Ghent and Ypres were growing fast and expanding their jurisdiction into the surrounding countryside.

This first industrial revolution, centered in Flanders, happened almost entirely because of the arrival from the Arab world of a new, horizontal loom, equipped with foot pedals to lift the warps. This innovation left the weaver's hands free to throw the shuttle back and forth, which made weaving much faster and more profitable and, above all, made possible the production of long pieces of cloth. Because of their centuries of experience in working wool, the Flemish were the best weavers in thirteenth-century Europe. Flemish cloth was sold everywhere in the known world, and its manufacturers went from the East Indies to the Baltic to obtain their dyes, and to the mines of the Middle East for the alum with which to fix the dyes so as to make their colors fast.

Part of the reason for this extraordinary Flemish success lay in a new sartorial fashion that spread across Europe in the late twelfth century. Woolen coats became fashionable among the rich, replacing the old linen garment of earlier years. Demand exceeded supply, and the Flemish had so many orders that they even had to import extra wool from England. The industry grew to enormous proportions, and the trades within began to be specialized. Most of the specialization occurred in those crafts dealing with the cloth after it had been woven. The brayer pounded the cloth to remove oil and dirt, the burler picked out loose knots and threads, the fuller matted and softened the cloth, the rower teased a nap up on the cloth, the shearman cut the nap smooth and the drawer mended any holes.

One thing above all ensured the continued health of the Flemish economy: the ability to push back the sea to create more and more land for the hundreds of thousands of sheep that represented the country's major natural resource. As early as the twelfth century, Flanders was already famous for its land-reclamation techniques. But in the thirteenth and fourteenth centuries, major climatic changes led to a rise in the sea level, and catastrophic floods overwhelmed the seawalls. On the night of November 19, 1421, in the area of Hollandse Waard, 105,000 acres were inundated, seventy-two villages destroyed and a hundred thousand people drowned. Then flooding grew worse, with major emergencies in 1468, 1526, 1530, 1532, 1551 and, most devastating of all, on All Saints' Day, 1570.

Not surprisingly, by the late sixteenth century — after a hundred years of flooding — the Dutch had the best hydraulic engineers in Europe. One of them was a man called Simon Stevin. Not much is known about his early life. Born in 1548, the illegitimate son of a wealthy Bruges couple, he began work in the Bruges city finance department; traveled to Poland, Russia and Norway while still in his twenties; and then studied at the University of Leiden. By the age of thirty-eight he was an established hydraulics engineer, writing about the structure of dams, sluices and locks, as well as the formation of sandbanks and other such subjects vital to the maintenance of polders: sections of waterlogged land around which a wall was built, the water removed, extra soil added and the whole thing left to dry out.

Stevin was also an expert on windmill design. Windmills were the principal machines used for lifting water, since the flatland lacked fast-flowing rivers where watermills might have been used. Stevin's contribution to the growing Dutch acreage was to design mills with larger scooping wheels that revolved more slowly and with transmission systems that used more efficient, conical, toothed wheels in their gearing systems. He also worked out the size and number of cogs on the gearwheels, so as to calculate the minimum wind pressure required on each foot of sail surface to lift the water to the necessary height. This information meant he could tell how much water would be raised by each turn of the sails.

Stevin's interests were, like many of his late Renaissance contemporaries, wide-ranging and included mathematics, astronomy, navigation, military science, music theory, civics, geography, house building and bookkeeping. Unlike them, however, Stevin believed in the value of the vernacular. He felt that Dutch was a good language for saying things clearly, and his own writing is lucid and entertaining. Unfortunately, few people outside the Netherlands understood Dutch, so much of Stevin's highly original work went unnoticed until it happened to be translated at a later date.

One of his books to reach general readership was *Interest Tables*, written in 1582, in which Stevin set out the rules of single and compound interest and gave tables for rapid calculation of discounts and annuities. The reason for this sudden concern for accountancy was that the Dutch economy was experiencing an unprecedented boom. Trade was growing fast, thanks to the development of a revolutionary new ship (called a *fluyt*)[112] which used pulley systems for raising and lowering sails so that crew requirements were minimal. The design also reduced the amount of deck housing, leaving plenty of space for cargo handling. And with its almost flat bottom, the *fluyt* had a large hold. All these characteristics combined to make the ship both efficient and cheap, and its shallow draft made it ideal for coastal and river transportation. By the late sixteenth century the Netherlands had virtually monopolized the business of delivery and re-export, bringing goods into Dutch seaports and distributing them, in *fluyts*, all over Europe.

Since the country was now swimming in money, it took the novel step of inventing an exchange bank so as to be able to lend the funds at interest. In 1585 Stevin boosted Dutch fortunes by taking the business of business a stage further. In a twenty-nine-page booklet with the rather underwhelming title of *The Tenth*, he changed a vital part of mathematical notation. In doing so, not only did he make

112 206 165

life easier for accountants, but he made possible at a stroke the immensely complex calculations that would be needed for the great astronomical discoveries of the next century. Stevin's arithmetic revolution introduced a radical new way to calculate.

Up till then, parts had been worked out using fractions that were written using integers (for example: $1\frac{2}{3}$). This method was extremely time-consuming when fractions were added (for example: $1\frac{19}{32} + 3\frac{12}{62} - 8\frac{46}{83} = ?$). The Hindu-Arabic numerical system was well established in Europe, and the shapes of numbers had settled more or less into their modern form. As the decimal point had been in use since medieval times and was well understood, Stevin's new idea was to apply decimals to fractions. Decimal fractions simplified calculation immediately, and the new system caught on like wildfire.

What was less immediately successful was Stevin's suggestion that the new system also be applied to coinage. In fact, nobody was to take any notice of this idea until late in the eighteenth century, in a distant country awash with as many kinds of money as there were places from which its many immigrants had come. When the United States was founded in 1782, it had no national currency. Americans used Spanish doubloons and dollars; Portuguese moidores and johannes; French livres, sous, pistoles and guineas; British pounds, shillings and pence; and various ducats, coppers, crowns, pieces-of-eight and pistareens. And all of these coins were worth different amounts in each of the different states and commonwealths.

The man who cleared up the coinage confusion was a one-legged aristocratic New York socialite who had spent time in France as U.S. minister during the time of the Reign of Terror, been elected to the Continental Congress, was assistant to his country's secretary of finance and rejoiced in the name of Gouverneur Morris. In 1782 he was directed by Congress to report on foreign currency circulating in the country. As a postscript to his report, he added his plan for the decimalization of American coinage. The idea was taken up by Thomas Jefferson,[113] who decided that the unit of currency should be the Spanish dollar (divided into a hundred cents), signified by a symbol that adapted the P and S from the Spanish word *pesos* and that was written as $. Since Jefferson was a great deal more famous than Morris, history forgot that it was the latter who had originally introduced the idea. But Morris was to make a mark on history every bit as big as the dollar.

113 80 53
113 213 172
113 295 261

In 1803 he told the surveyor-general of New York State, Simeon DeWitt, about his idea for "tapping the waters of Lake Erie . . . and leading its waters in an artificial river, directly across the country to the Hudson River." In spite of great general enthusiasm for the concept, Jefferson thought it was a hundred years too early and would not release federal funds until, in 1810, commercial pressure grew too intense to ignore. The mayor of New York, DeWitt Clinton, held massive rallies to express public support for the idea. But thanks above all to the continual efforts of Morris himself, in 1816 the Erie Canal Bill finally passed through Congress and work began on it the following year.

When it was completed seven years later, the canal ran 363 miles from Lake Erie to the Hudson and incorporated eighty-seven locks, including five pairs of double locks. The canal was a marvel of the engineering world, although, in spite of its immense length, its dimensions were modest: only forty feet wide at the surface and twenty-eight feet wide at the bottom, with a water depth of only four feet. Possibly for this reason, it was often referred to as "Clinton's Ditch."

On opening day, October 25, 1825, festivities were organized along the entire length of the canal and featured a triumphal procession of boats hauled by horses down the canal to New York City, where two kegs of Lake Erie water were ceremoniously poured into the Atlantic to symbolize this new union. Once the canal was in business, Morris's predictions proved correct. Freight rates dropped to one-tenth what they had been before the canal, and business boomed all along the towpath. In the first ten years revenue repaid the costs of construction, guaranteed the supremacy of New York as prime entry port to America, carried twelve hundred immigrants a day to Detroit and turned Chicago from a village into a city. In 1852, when more than three hundred thousand immigrants arrived in New York, most of them went west on the canal. Celebrated in song and story, the Erie Canal was immortalized by the poet Philip Freneau even while it was still being dug:

> By hearts of oak and hands of toil
> The Spade inverts the rugged soil
> A work, that may remain secure
> While suns exist and Moons endure.

Freneau's optimism ignored the lightning strike of change. Only twenty years after the opening, alongside the great canal, construction work began on a radically new way of taking immigrants west that would make the Erie Canal (and every other canal) instantly obsolete. It was a railroad, and it beat canals hands-down because it provided a more direct route, was cheaper to build over rugged terrain, didn't need a constant supply of water, was less expensive to maintain and, most important of all, was the first-ever form of freight transportation to move faster than horse-drawn barges.

In 1845 the New York & Erie Railroad first linked Lake Erie and the Hudson River, and its arrival brought into being modern business methods. One of the first superintendents on the Erie Railroad, Charles Minot, was the man who solved the early problem plaguing the new iron-horse lines: delays caused by single-track trains having to wait in sidings so that they could pass each other safely. With hindsight, the solution should have been obvious. In 1851 Minot used the telegraph,[114] whose wires were strung alongside the rails, to signal the oncoming train when to wait and when to proceed. This practice soon gave rise to a regularized time-table code-system that would not be superseded for another thirty years.

114 30 29
114 235 197
114 275 237

In 1854 the new superintendent, Daniel McCallum, found that railroad administration was in chaos, so he invented a pyramidical organizational chart to show in what way detailed and accurate management information should flow through the system. People were so impressed with his chart that it went on sale for a dollar, was mentioned in British parliamentary discussions and appeared in the *Atlantic Monthly*. McCallum also worked out hourly, daily, weekly and monthly reporting procedures and got managers to use the telegraph to maintain contact with their staffs.

Even as McCallum was taking his post, the new Pennsylvania Railroad was inaugurated. Its president, J. Edgar Thomson, saw that the future profitability of the railroads lay primarily in picking up freight from its point of origin and delivering it direct to its final destination. So he developed the concept of line-and-staff management and the "divisional" type of organization. These systems were so effective that the Pennsylvania became the largest business enterprise in America — by 1880 employing fifty thousand people.

Next, the Baltimore & Ohio synthesized McCallum's and

Thomson's work into an organization consisting of separate administrative, finance, operations and legal departments. Finally, Albert Fink, who in 1869 was vice-president of the Louisville & Nashville, worked out how to find the most important management information of all: the cost of operation per ton-mile. To do so, he reordered all financial and statistical data compiled by the accounting and transportation departments, and reorganized existing accounts according to the nature of their costs (instead of according to which department carried out the functions). Then Fink put all the data into four general categories so he could compare how different stations were run and find out why their costs differed.

All these improvements came together in the American "divisional" type of organization. Local department or division heads were delegated authority to control and coordinate traffic, giving the railroads flexibility of control over their operations, some of which took place hundreds of miles away from headquarters. The railroad business was also the first to use full-time professional managers, working in a formal managerial hierarchy, running large-scale operations over vast distances and handling thousands of customers as well as hundreds of types of goods and services. The railroads had invented modern corporate structure.

Almost immediately, they were imitated by one particular commercial entity which owed its existence to the railroads and whose business procedures mirrored theirs. It was the department store; and like the railroads, it dealt with high-volume inventories, massive turnover, thousands of customers buying at hundreds of different points of sale, a wide variety of commodities moving at high speed through the system, massive capital outlay, low profit margins and the requirement for day-to-day information on inventory and cash flow.

The new department stores revolutionized commerce. They used telegraphs, railroads, steamships and postal services, and used the same "departmental" administrative structure as the railroads. In the 1870s stores like Marshall Field in Chicago and Stewart's in New York worked mainly wholesale, with retail trading representing only about 15 percent of their turnover. But by the 1880s, with improved streets, better urban transportation,[115] elevated railways and the rapid rise of city populations, the retail side took over.

Shopping in these new commercial palaces — with their chandeliers, marble floors, neoclassical doorways, plate-glass windows,

The clipper *Taeping* leading the *Ariel* in the great tea race of 1866. The clippers' extraordinary speed was due to their new sail plan. The three traditional booms mounted on three masts now became four, and sails were split so as to provide a more versatile sail surface, as well as easier management in foul weather.

WHERE THE GOLD COMES FROM.

An 1849 painting of the California gold diggings. Prospecting was a relatively easy task because climatic and geological circumstances made finding gold fairly simple, since erosion, prehistoric glacier movement and ancient, gold-bearing riverbeds thrust to the surface by volcanic activity put gold literally within reach of anybody with a pan or shovel.

The V-2 on the launch pad at Peenemünde. The four large external aerodynamic vanes were attached to four graphite directional vanes projecting directly into the rocket exhaust. After the war, nearly one hundred V-2s were sequestered by American forces and taken to White Sands, New Mexico, where they provided much data later used to build U.S. moon rockets.

A schlieren photograph of the shock waves created by a projectile in a wind tunnel. This illustrates the kind of phenomenon observed in experiments carried out by Ernst Mach in 1889, during which he established the speed of sound as "Mach 1." The schlieren technique was originally developed by August Topler in 1864 to "see" sound waves.

F. T. ARCHIVE

The Union Pacific's Dale Creek railroad bridge in Wyoming, a typical cause of the Great American Wood Famine. Built of Michigan pine in 1868, it measured 700 feet in length and rose 126 feet above the streambed. At the high point of American railroad construction, thirty million trees were being felled each season.

BY PERMISSION OF THE PRESIDENT AND COUNCIL OF THE ROYAL SOCIETY

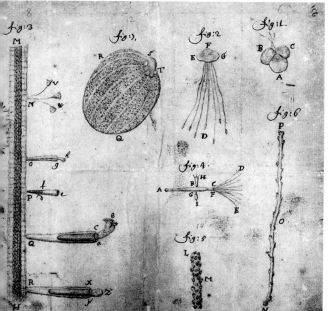

One of the drawings that shocked the eighteenth-century intellectual world. The page is part of Leeuwenhoek's letter to the Royal Society on Christmas Day, 1702. It shows a series of studies of organisms associated with duckweed, in particular the rotifers (fig. 3: Q, R). The drawings are so accurate that they would not be out of place in a modern introductory biology text.

Part of a report on the American Civil War, from the *Illustrated London News* of April 12, 1862, and featuring an illustration of the famous encounter between John Ericsson's *Monitor* (almost submerged, at center) and the southern ironclad, the *Merrimac* (carrying a flag, on the left). After four hours the "invincible" *Merrimac* withdrew.

A contemporary manuscript illustration of the battle of Crécy, which marked the zenith of the "English" longbow. Earlier on the day of battle, the English had unstrung their bows because of rain, but French crossbowmen were unable to do so, and their strings were damaged. Interestingly, the left margin shows a cannon, the new weapon that would replace the longbow and change the face of war.

The October 25, 1825, opening of the Erie Canal by Governor DeWitt Clinton, who ceremoniously pours a keg of Lake Erie water into the Atlantic at Sandy Hook, New York. The canal ran from Buffalo to Albany, was dubbed "the eighth wonder of the world" and boasted eighty-three locks.

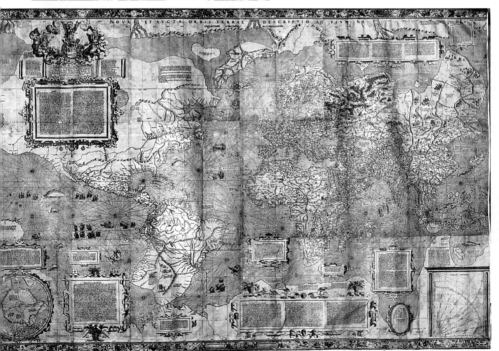

One of the most influential maps of all time, Gerard Mercator's wall-chart of the world as it was known in 1569. The map shows the first lines of latitude and longitude, projected to allow a ship's course to be plotted as a straight line with a ruler. Note (top left) how Mercator imagined the Northwest Passage linking the Atlantic and Pacific.

Flensing a whale (removing the blubber) in 1574. By 1619 the Dutch had established a permanent whaling community on Spitsbergen. Called "Blubber-town," at the peak of the season its population rose to more than a thousand men, working up to twenty ships. Thanks to the low cost of borrowing in Holland, the Dutch led the whaling industry for decades.

A German print (every country in Europe published one) of the first manned flight in a Montgolfier hot air balloon in November 1783. Of the two aeronauts on board, Pilatre de Rozier was later to die during another similar flight. The Montgolfier brothers were planning to develop a dirigible version of their balloon when the French Revolution cut short their attempts.

A design from a nineteenth-century Kashmir shawlmaker's pattern book, showing the kind of patterns which became popular all over Europe and America after they were woven in Paisley, Scotland. Patterns this complicated could be safely mass-produced only on an automated loom.

Drawings (from Abraham Rees's 1820 *Cyclopedia*) based on Blumenbach's division of the human race into different racial groups. Blumenbach based his classification (which established the science of anthropology) on form of skull, skin color, hair and body structure. He later went on to write the groundbreaking *Handbook of Comparative Anatomy*.

The famous "Colonel" Edwin Drake (on right) at Drake's Well in Pennsylvania. The tall wooden building housed the stack of tubes and drills, often operated by salt drillers who had experience with oil seepage into their wells in West Virginia, Kentucky, and Ohio. Drake's Well was not, as he claimed, the first. And Drake himself was not a colonel, but a train conductor.

Double Cabinet Lavatories with Marble Slabs.

The epitome of Victorian sanitary go taste, produced by the Crapper compa (considered to have invented the first flu lavatory, in sixteenth-century Elizabeth England). The price of this unit would ha been the equivalent of a sewer worke wages for more than a year. By 18 copiously illustrated catalogues made t technology available by mail order.

No. 820

No. 820. Lavatory for Two Persons, comprising Mahogany or Walnut Stand and High Mirror Back, with Hand-painted Tiles ; 1¼ in. White Veined Marble Top, Moulded Edges and Soap Sinkings, 20 in. D Basins, Hot and Cold Valves and Quick Waste Arrangement.

Size of Slab ... ... 5 ft. 0 in. × 2 ft. 0 in.
Price Complete £44 7 6

If with Plated Fittings, Extra 15/-

rotundas and galleries — became a cultural experience. The kind of merchandise on sale ranged from the dry goods with which many of these stores had started, to luxury items like jewelry, glassware, silver, clocks, books, hosiery, handkerchiefs, gloves, artificial flowers, feathers and furniture.

The new "convenience" shopping took place in emporia whose staffs often were larger than the population of many American cities. Shoppers were referred to as "guests" and pampered with beauty salons, restaurants, live musicians, restrooms and post offices, as well as delivery and repair services. The cost to the department stores of services, outlay on buildings and furnishings, size of the staff and the low profit margins meant that it was soon essential to find new ways of persuading customers to spend more money.

The man who showed them how was Henry P. Crowell of Ravenna, Ohio. In 1881 he bought a bankrupt mill for $25,000 and went into the business of convincing people to consume what previously only poverty-stricken Scotsmen, Germans and horses had eaten. Crowell persuaded people to like oats. He did so with the first real industrial advertising campaign, every element of which was specifically targeted. To begin with, he chose as the brand image for his product the smiling face of a Quaker. People trusted Quakers because they had a reputation for fair dealing and a clean, honest lifestyle. So Crowell called his product Quaker Oats. The Quaker brand was the first to combine all the various kinds of advertising in use at the time. Crowell's print ads emphasized the food's nutritional value, with lines like "The one overwhelmingly important thing in this life is health" and "The cereal-eating nations of the world can endure more physical toil than the meat-eating nations." Crowell introduced free gifts, money-back guarantees, box-top coupons, testimonials from famous people and scientific endorsements. He even created the "Quaker Train" that toured the West with free samples.

But perhaps Crowell's greatest contribution to modern merchandising was his use of packaging. The oats came prepackaged in a box that appealed to housewives because the packaging ensured purity of product and quality control and offered standard-weight value for money. The packaging also impressed the store owners and distributors because it was easy to transport and display. Crowell had kicked off modern marketing, and he was astonishingly successful at it.

Now that the consumer had been persuaded to buy, it remained only to persuade the factory workers to produce. In 1924 came a major study of worker motivation that was to affect every aspect of production and personnel management from then on. The survey was conducted among the twenty-nine thousand staff of the Western Electric Company's Hawthorn works in Chicago. Over a period of five years, attempts were made to learn how output was affected by the factory environment. The investigators began by introducing variations in the levels of illumination, room temperature, humidity, hours of work, periods of rest, food consumed, hours of sleep at night, length of working week and holiday periods, as well as changes in general working practices.

For the first two years the results of the experiments seemed to prove the premise that the major factor affecting output was fatigue, but then anomalies began to appear. When one group of workers was told that conditions were going to be improved (but in fact conditions were left unaltered), their output improved. And on other occasions, after better ambient conditions had improved output and workers had been informed of this fact, conditions would be secretly returned to what they had been before — but output still remained high. It was only after a series of exhaustive interviews that researchers discovered what was really happening. Simply involving the workforce in the preparations for tests had improved personnel relations so much that this alone had been sufficient to improve performance. Also, on several occasions the mere expectation of improved conditions had led to improved performance, even when no such improvement had taken place. This was the first example of what became generally known as the placebo (or Hawthorn) effect. The belief that improvements are taking place is sufficient to alter the physical and emotional state of the worker.

These findings confirmed research being carried out by Walter Cannon, professor of physiology at Harvard. Some years earlier he had been excited by the new, X-ray[116] view of the body's swallowing mechanisms and had started giving patients with gastrointestinal problems a meal mixed with barium sulfate (now known as the barium meal). When viewed with X-rays, the movement of the barium meal along the alimentary tract revealed the presence of stomach waves. This led Cannon to investigate the phenomenon of hunger

116 39 *31*
116 226 *187*

and to discover that it is generated by stomach contractions and associated dryness of the mouth.

Cannon had also seen that these stomach waves of laboratory test animals abruptly ceased if the animal were frightened or disturbed, so he switched his attention to the effect of emotional state on physical systems. When Cannon removed an animal's sympathetic nervous system, it was unable to resist changes in the physical state. This condition was accompanied by the disappearance of a substance which normally appeared in the bloodstream on such occasions. Cannon realized that this was probably a chemical messenger of some kind that would normally help the body to respond appropriately to different physical or emotional states.

Cannon had seen that physiological reactions often continue long after the chemical stimulus causing them is removed, and from this he was able to identify the role of adrenaline in enabling the body to handle stress. In times when the appropriate response to a situation is flight or fight, the adrenal gland causes massive increases in blood sugar to be released. Blood then shifts from the abdomen (where it is normally needed for digestive processes) to the heart, lungs and limbs. In the 1920s Cannon suggested that higher animals had a complex repertoire of self-regulatory mechanisms involved in keeping the body stable under a wide range of physical and emotional circumstances. In a groundbreaking book called *The Wisdom of the Body*, published in 1932, Cannon coined the phrase "homeostasis" to describe this process of internal balance-keeping.

Cannon's favorite collaborator over the last fifteen years of his life was a physiologist at the Harvard Medical School called Arturo Rosenblueth. Later, during the first years of World War II, Rosenblueth discussed Cannon's work with Norbert Wiener, a mathematics prodigy who was professor of math at MIT. At the time Wiener[117] and his assistant, Julian Bigelow, were working on 117 122 92 a problem related to anti-aircraft artillery, and they were keen to talk about homeostasis and how it operated in humans. They were particularly interested in the way information from the eyes and the body-position sensors fed signals into the brain so as to make it possible for a person to pick up a glass of water or a pencil without over- or undershooting the goal, at each end of the event.

From these conversations with Rosenblueth, Wiener formulated a general concept which he called cybernetics (based on the Greek

word for "steering"), which he and Bigelow applied to their anti-
aircraft[118] artillery work. They expressed this cybernetic concept
in mathematical algorithms that made it possible to process incomplete radar data about the previous track of a target, so as to calculate the target's most likely future position. With this data, artillery
could be aimed at where the target would probably be when the
shells arrived.

In 1944 the new system first appeared in the form of the M-9
predictor; and during the first few weeks of its use against incoming
V-1 missile targets along the English Channel, it was a resounding
success. In the first week of the final month of German missile[119]
attacks, 24 percent of the targets were destroyed. On the last day of
missile raids, in which 104 rockets were launched, 68 were destroyed
by Wiener's cybernetically controlled guns. Britain was saved.

After the war cybernetics became a major feature of computing
and automation of all kinds, and the feedback principle was soon
employed in many types of factory machinery. One of its more dramatic modern uses has been in inertial navigation systems. Inertial
gyroscopes[120] sense the direction in which an aircraft or a missile
is traveling, and accelerometers react to any changes in velocity.
Both systems work together with electric motors that sense any
change in these instruments' state because of changes in orientation
or speed, and these data are used as feedback to return the instruments to their initial state thousands of times a second. The electric
charge required to make this adjustment depends on how far from
initial the instruments have moved. Integrated with a clock, these
data show where the vehicle is at any time.

So thanks to the longbow, modern warfare can be conducted with
immensely flexible missile systems that use cybernetic feedback to
compensate for changes in the battlefield environment — such as the
atmospheric state or the height of terrain below — and to guide
themselves unerringly across hundreds of miles. Onboard inertial
guidance systems ensure that modern missiles can hit their distant
targets with a circular error probability of only a few feet. A degree
of accuracy that even Robin Hood would have been proud of.

Modern, hi-tech warfare is fought at a distance, with detection
systems and countermeasures coming into play, on many occasions
long before individual humans are directly involved. These tactics
owe their existence to one of history's all-too-frequent accidents. . . .

7 High Time

MUCH modern technology is so pervasive that it often goes entirely unnoticed. People speak of "transparency" when describing information systems that interact with others as easily as if there were nothing between them. The goal is to make innovations so "user-friendly" that most of the time we're unaware of their presence. Plastic wrap is perhaps one of the best examples of this. It is essential, pervasive, easy to use and (literally) transparent. And like almost every other piece of technology, anywhere and anytime, it came into existence by accident.

Plastic wrap first came to general attention early in World War II. At the time, German bombers were conducting nighttime raids against England almost at will.[121] What little opposition the 121 119 90 Germans encountered would be a few fighters vectored against them by controllers working under the considerable limitations of the early-warning radar of the time. So warning was not very early. In addition, the radar, which was positioned along the south and east coasts of England, used long-wavelength signals requiring large aerials that were an easy target. Therefore, the British were keen to find ways of transmitting radar signals from smaller aerials. Either that, or the war was going to be over almost as soon as it started.

What they didn't know was that the problem had already been solved, thanks to an accident on March 24, 1933, when researchers in the British chemical company I.C.I. were using special glass vessels (known as bombs) to make dyestuffs under extremely high pressure. One of the "bombs" exploded and some waxy white residue was noticed on the vessel's inlet tube. Then the same thing happened several more times. Finally, in 1935, the nature of the residue was recognized as a molecule which was a polymer (meaning "made up of many parts"), water resistant and (most important) an excellent electrical insulator. The material was named polyethylene and, once in regular production, was made by machines that blew films of it, much in the same way a wire loop can be used to blow soap bubbles.

Polyethylene was already being manufactured just before the start of the war, but its value to radar came to light only when some of it was offered to researchers in the atomic weapons research unit, who were looking for a good insulator. It was this insulating property that would give Britain a decisive military advantage by making it possible to construct much smaller, higher-frequency radar sets,[122] whose relatively weak signals would no longer leak away. Higher frequency meant better definition, more accurate signal returns from targets, and radar equipment small enough to be carried on ships and planes. By 1943 British night-fighters using three-centimeter-wavelength radar were shooting down German bombers, and the British fleet was successfully engaging enemy warships in the dark. As a result, U-boat losses forced the Germans to call off the Battle of the Atlantic.

One of the most frequently used ways to make polyethylene, as noted, is by blowing bubbles of it. Plastic makes bubbles because of the way its very large molecules join together in long chains of great strength and durability. This is why plastic can be made to behave like soap, which for the same reasons can also display great durability. The nineteenth-century Scottish scientist James Dewar once kept a soap bubble "up" for three years.

Plastic and soap behave the same way because both are "colloid" materials — which means that they diffuse across membranes easily. In the case of soap, the molecules group readily into extremely large clumps called micelles. These turn out to be a key element in the process by which soap makes things clean. Soap coats dirt and grease particles with molecules that make the particles attractive to water, so they loosen a little from the fabric (or your skin), become globular in shape and then detach. Once these particles are floating free, the micelles wrap around them and prevent them from reattaching. So the object (or your skin) gets clean and stays that way.

It's not often that somebody becomes a national hero, courted by great artists and showered with honors all because of soap, but such was the case with M. E. Chevreul. He began his professional life as a young French chemist living in Paris, where he became the first person to find out how soap worked. In 1811 he had started to study dyes and the complex mixture of plant oils and resins they came from. This study led him to look at fats, and in due course he discovered fatty acids. In 1823 he published his first major report, stating that soaps consist only of fatty acids and a base. Chevreul listed the kind of fatty acids that would be suitable for saponification and turned soap making into a science. And since fatty acids could also be used to make candles, he lit the place up, too. Thanks to Chevreul, it was a bright, clean-smelling world; everybody who mattered knew it; and it made Chevreul instantly famous. When he died at the age of 102, France declared a national day of mourning.

Chevreul had one other reason for being interested in soap. He was the director of dyeing at the already world-famous Gobelins[123] tapestry factory in Paris, and dyes behave like soap when they get onto cloth. Chevreul was trying to make natural dyes more intense and to make sure they didn't wash out easily. While working on these problems, he discovered that the intensity of a color related less to the strength of its pigmentation than to the hue of the colors placed next to it. This juxtaposition affected how the eye saw the color. Chevreul's "law of simultaneous contrast" led him to invent an extraordinary new color tool. He took the three primary colors — red, blue and green — interspersed them with twenty-three color mixtures and got a chromatic circle of seventy-two colors. Then he toned each color by adding black or white, thereby creating the 15,000-tone chromatic circle used by all dyers ever since.

But Chevreul's placement of color for effect did much more than help the textile industry. It also changed the world of art by triggering the French "scientific" impressionist movement. Painters like Seurat, Signac and Pissarro used Chevreul's new law of contrast in their work. They placed spots of different colors next to each other, to create the impression of a third color, and in doing so achieved the distinctive, shimmering effect for which impressionism is famous. Perhaps the greatest (and most valuable) example of

Chevreul's ideas can be seen in Seurat's *Un Dimanche à la Grande Jatte.*

One of the reasons for Chevreul's original obsession with color contrast might have been the urgent demand for more hues in the tapestries being produced at the Gobelins factory where he worked. Fifty years earlier, the factory's director, François Boucher, had been a great fan of objets d'art from China. From the mid–eighteenth century, French taste had been captured by the decorative art being imported from the mysterious East, and *chinoiserie*[124] was all the rage. The new fashion gave rise to a much more complex tapestry style, in which the weavers used light but varied materials, intermingling threads of silk, linen, wool and cotton to create greater subtlety of effect. This complexity was why a greater number of hues was needed, inspiring Chevreul to produce his 15,000-tone color wheel.

124 242 205

The French mania for things from China became so extreme that a special section of the Gobelins factory was set up by the government in 1762. Called the *Ouvrage de la Chine* (the China-work factory), its purpose was to make Chinese and Japanese fakes — in particular, lacquer furniture, by far the most eagerly sought product. Well-made fakes would satisfy public demand and reduce expenditure of the country's meager reserves of gold (which was all the Chinese would accept in payment for their exports).[125]

125 95 66

Lacquer furniture had first turned up from Japan in 1610, when nine decorated chests arrived in Holland on board the Dutch East India Company ship *The Red Lion.* Right from the start, demand and rarity value pushed prices through the roof, and there were so many potential customers for lacquerwork that the newly formed company was desperate to find another way to get to the source of supply. The problem was that the two established routes to China and Japan (round the bottom of Africa and across the Indian Ocean, or round the bottom of South America and across the Pacific) were controlled by the Portuguese and Spanish, respectively.

So the Dutch were looking for an alternative and, if possible, shorter way to go. Ever since the late fifteenth century, explorers like the Cabot family had tried (and failed) to find the Northwest Passage, through the polar seas north of Canada and Alaska, to the Pacific. Everybody was convinced the route was there, because nobody had ever gone far enough north to discover that the permanent polar ice cap would make such an attempt difficult, if not impossible. Henry Hudson, an English explorer, was encouraged to try his luck. He knew John Smith (of Pocahontas fame) and may have thought that Smith's descriptions of the Great Lakes referred to the Pacific; so when he was in Holland in 1607 and the Dutch East India Company asked him to try to find the passage, he was already half-persuaded.

The voyage undertaken for his Dutch masters took him to Greenland and then Spitsbergen, and back to Greenland. Every time he attempted to get farther north, he was blocked by the ice cap. Finally, he gave up and went home, but not before discovering that there were whales in plenty to be found off the coast of Spitsbergen. By 1619 the Dutch had set up a whaling[126] industry on nearby Amsterdam Island. At this time, the most financially rewarding thing afloat (this side of a Spanish galleon) was a whale. Whalebone went into brushes, handles, sieves, packing cloth, crossbows, bed bottoms, carriage backs and sofa frames, and was also used as stays for corsets, ruffs and dresses. Whale blubber made candles and soap, and whale oil fueled lamps. Whaling generated 500 percent profit on investment and, given the low cost of Dutch borrowing, Holland was soon running the biggest whaling industry in the Atlantic. The whales off Spitsbergen were so profitable because the waters were so cold that the animals were covered in lots of blubber. Holland was to continue making profits from whaling for over three hundred years. So Hudson's voyage hadn't been a total failure after all.

The man who had originally persuaded Hudson that the route he took was feasible was the cartographer Pieter Platvoet, better known by his pen name, Plancius. Plancius was the one who had persuaded the Dutch authorities to set up the Dutch East India Company in the first place. He was also a proponent of the northern route, and his views were taken seriously because he had been one of the most brilliant pupils of the great cartographer Mercator. Mercator had become famous all over the continent because he was published by the greatest printer in Europe at the time, Christophe Plantin.[127]

At his Antwerp printing house (which is still there today) Plantin ran twenty-two presses, as well as a profitable sideline in French underwear, wine, fancy leather and mirrors that he sold through an

126 61 45

127 204 164

agency with branches everywhere from Sweden to Algeria. But where Plantin made his fortune was in holy books, thanks to Philip II of Spain and the Council of Trent. By the mid–sixteenth century, because of Luther, the faithful were deserting the Catholic Church in droves and defecting to Protestantism. So in 1656 Rome decided to do something about the problem and convened a meeting of bishops in the northern Italian city of Trent. The meeting lasted several decades and essentially resulted in four decisions: to create more effective propaganda by commissioning what is known today as baroque art; to give the Jesuit order authority to crack down on dissidence and set up colleges all over Europe; to standardize all liturgical texts so that everybody was conducting the same services and saying the same prayers; and, last but not least, to establish an index of prohibited books.

It was the standardized liturgical texts that would make Plantin rich, because the council decree inspired Philip II of Spain not only to commission him to print forty thousand copies of new, approved editions of missals and breviaries but also to agree to Plantin's proposal for a totally new kind of "scientific" Bible. Work on it began in 1568, under the watchful eye of Philip's representative, Arias Montano. The job took eleven years and in the long run was to have an effect far beyond that of the Bible itself. The "scientific" nature of the new Bible lay in its appendices, which dealt with subjects like biblical coinage, genealogy of the prophets, Judaic medicine, Aramaic weights and measures, dictionaries and grammar of the biblical languages, the flora and fauna of Israel and maps of the Holy Land.

Plantin assembled a number of experts to work on these appendices; once they had done their work, they used their newfound experience with printing to set themselves and their subjects up as autonomous intellectual disciplines. We refer to the eventual effect of this "knowledge fallout" as the Scientific Revolution, which began when these new experts brought their editorial skills to the critical examination and investigation of the data in classical Greek and Roman texts dealing with subjects like botany, medicine and cartography.

As far as the new index of prohibited books went, ironically one of the first texts put on the list was one that had originally been commissioned by Rome itself. The problem was this: due to the fact that the accepted model of the solar system at the time was Aristotelian (which incorrectly placed the Earth at the center of the solar system), the date of the festival of Easter (which depended on calculations involving the relative positions of the Sun and Moon) was wrong, by an unknown amount. However, since the proper observance of holy days was essential to an individual's salvation, the correct date for Easter had to be found. So the church ordered a Polish astronomer and cleric named Mikołaj Kopernik (better known as Copernicus[128]) to sort it out. He did so by removing the Earth from its central position and placing it (and the other planets) in orbit round the Sun. But since this placement removed humankind from its biblically sanctioned central position in the cosmos, Copernicus's book, published in 1523, was now defined as heretical and placed on the prohibited list.

128 246 *210*

And then in 1610 Galileo Galilei,[129] an Italian professor of math, put the astronomical cat among the theological pigeons with drawings of what he had seen through an amazing new device called a "looker," invented in 1608 by an obscure Dutch eyeglass maker called Hans Lippershey[130] (and turned down for use on the battlefield by his patron, Prince Maurice of Nassau, on the grounds that he had really wanted binoculars). By 1609 the telescope was being manufactured in Paris and London; Galileo heard about it and made one for himself. Galileo used his new instrument to do dangerously heretical things. He observed phenomena like mountains on the Moon and spots on the Sun (however, the church said the heavenly bodies were supposed to be perfect, featureless globes). Worst of all, Galileo saw moons circling Jupiter (though the church said everything circled the Earth). Finally, he compounded the felony by observing the transit of Venus across the Sun. This was only possible if the solar system were, in fact, Sun-centered. Galileo's publication of his findings, called *The Starry Messenger*, caused a sensation.

129 168 *133*
130 158 *126*
130 247 *210*

In 1611 Galileo went to Rome to show the Jesuits his discoveries, impressing them so much that the Roman College (the Jesuit intellectual powerhouse) soon became a center for astronomical research. Nonetheless, the news Galileo was spreading did go directly counter to church teaching. Following Jesuit advice, Rome offered to impose only a temporary delay in the publication of Galileo's new work (called *The Dialogue of Two Worlds*). The argument was that it would take time to adapt church teachings to the new reality; and when these steps had been taken, Galileo would be free to publish. But the man was intransigent and published the book in defiance of Rome. In 1632 he paid for his stubbornness with lifelong house arrest and prohibition from any further publication.

The supreme irony was that the man who upgraded the telescope (and saw much more than Galileo ever had) was a German Jesuit, a contemporary of Galileo's named Christophe Scheiner, professor of Hebrew and math at Ingolstadt. He, too, observed sunspots but believed them to be tiny planets. Scheiner's telescope improved on Galileo's (which had been only a foot long, with one convex lens and one concave lens, and showed about only half the face of the Moon). Scheiner used a strong convex lens at the eyepiece and a weaker one at the other end. This gave a twenty-four-inch focal length and a sharp, though inverted, image. Increasing the focal length increased magnification to a virtually unlimited extent. This improvement eventually led to telescopes[131] as long as 150 feet, suspended by ropes and pulleys. Though these long instruments moved in the slightest breeze, they made possible seventeenth-century discoveries such as Saturn's moons[132] and rings, the "canals" on Mars (first seen by Scheiner) and Jupiter's belts. At one point, a telescope one thousand feet long was planned by a French astronomer so that he could see animals on the Moon.

These new telescopes also made life safer for sailors, when people like the French astronomer Cassini obtained precise fixes on the times and positions of the moons of Jupiter and made these refinements available to navigators' star tables. The tables were then used to measure the position of a heavenly body, in order to compare its position with what it would have been at the same time back at base. The difference in position and time told the navigators how far east or west they were. However, a highly precise star fix was essential, for the simple reason that the Earth turns fifteen miles in a minute, so a star position which was inaccurate by only one minute of arc gave a position that was wrong by fifteen miles. And in bad visibility that margin could mean missing your destination.

Things were complicated by a French expedition to Cayenne, near the equator in South America. The expedition's astronomers noticed that the pendulum on their clock swung a little less than it had done in Paris. They believed this difference meant it weighed less; and if this were so, the Earth was not a perfect sphere. On a perfect sphere, gravity would be the same everywhere. The prevalent theory was that columns of matter stretched from the Earth's center to the surface, and that equatorial columns were less dense and therefore created less gravity. This theory explained why the pendulum weighed less and swung less: if the columns were less dense at the equator, they had to be more stretched out — which means that the planetary diameter had to be greater at the equator than at the poles.

Traditional believers in a spherical Earth found this premise hard to take, and argument raged. Newton and the English maintained that the Earth was an oblate spheroid, flattened at the poles. The French did not. This was more than an academic dispute. If the shape of the Earth were wrong, so too were the maps based on it, and the navigation based on the maps. Things were given new urgency when inaccurate navigation by the notably named Sir Cloudesley Shovell,[133] Admiral of the English Fleet, caused him to hit the rocks off the Isles of Scilly one foggy night on his return from Gibraltar in 1714, thereby losing all his ships, two thousand crew and his own life.

133 283 248

There was only one way to resolve the dispute (and the navigational problems), and that was to measure one degree on the surface of the Earth at the equator and one in the far north, and see if there was any difference. In 1735 a French scientist, La Condamine, was dispatched to Peru. A year later a second group, headed by Maupertuis, left for Lapland. Maupertuis's technique for measuring one degree of arc was to fix a point on the Earth by taking a very precise star fix at a very precise time. Then he moved north across country until the position of the same star, at the same time of night, was one degree different. At this point he noted the distance he had traveled away from his earlier fix.

This sounds easy, but was in fact extremely difficult and, above all, required extraordinarily precise measurements. Before he left, Maupertuis had gone to London (the hi-tech instrumentation center at the time) and bought a nine-foot telescope whose aim was controlled very precisely by a micrometer screw in its base and carried silver-wire crosshairs on the lens. The telescope was also spring-mounted, to compensate for the effects on the instrument of the extremes of temperature that it would experience over the year of work. Maupertuis also bought the latest pendulum clock.

These instruments enabled the expedition to discover that the degree of latitude in Lapland was, after all, longer than a degree measured in France. Then in 1738, the French expedition to Peru returned with news of an even shorter equatorial degree. The British had been right after all. Accurate maps could now be made.

The British had another reason to be pleased by how things turned out, since the instruments that had done such good service in Lapland had been made by George Graham, England's greatest instrument maker. He was (and still is) best known for his clocks because he had come up with a clever invention to solve the problem, bedeviling all pendulum clocks, of tooth-stick. This problem occurred because the basic energy to drive a clock came from a weight suspended on a cord wound round a driveshaft. As the weight hanging on the cord slowly fell, the cord pulled the driveshaft round. This in turn rotated a toothed wheel set on the shaft. However, the rotation of this wheel was stopped and then released in small increments, thanks to two inward-pointing flat metal spikes (an arrangement called pallets). These pallets were mounted at each end of an upside-down, arc-shaped band of metal, whose top center was attached to the pendulum. As the pendulum swung first one way, then the other, each end of the arc would move laterally in and out, causing each pallet in turn to swing in and engage the wheel teeth. As the pendulum swung, one of the pallets would swing out and away, and the toothed wheel would be released to turn (because the driveshaft was being pulled round by the weight). At this point the other pallet would swing in to trap the next wheel tooth, and hold it for a moment, before the movement of the pendulum once again moved the other way and the pallet retreated, releasing the wheel, and so on.

The problem for precision fanatics (like the French degree-checkers) was that pendulums could swing erratically. When this happened, the pallet would engage or disengage clumsily, causing a minor delay in the turn of the wheel. To star-fixers this was an unacceptable imprecision, because the slightest inaccuracy in timing would affect their calculations.

George Graham's improvement made all the difference. All he really did was make sure the pallets did not stick in the wheel teeth,

even if the pendulum swings were erratic. He did this by rede-
signing the pallets in an arrangement that became known as deadbeat escapement. The redesign involved shaping the sides of the
pallets in such a way that the wheel tooth moved round, to rest
against a flat pallet edge, which immobilized it. At the next pendulum swing, the curved edge on the other side of the pallet pushed
the wheel tooth away from it. So the wheel was locked, each time,
at rest. Graham's escapement meant that the clock hands moved
very smoothly and exactly, thus making precise time measurement
easier. It was this facility that allowed Maupertuis to make extremely fine calculations when fixing his star positions.

The deadbeat escapement raised expectations of accuracy in all
quarters, including those of the municipal authorities in the rapidly
growing cities of the industrial nineteenth century. Fancy new
town-hall buildings often incorporated clock towers, which posed a
new problem. Bat droppings, grease and dirt would accumulate on
the drivewheel, or ice and snow would pile up on the external clock-
face hands. Either condition would affect the drivewheel enough
to make its movement less than smooth, in turn affecting how the
pendulum and the pallets would interact with it. The result would
be inaccurate timekeeping.

Municipal annoyance at this problem became a matter of British
national importance when Parliament ordered the building of a
clock tower containing a clock which, well before the tower was
even completed, was already being advertised as the most accurate
clock in the British Empire (and therefore, of course, the world).
The difficulty was resolved by a lawyer named E. B. Denison (later
ennobled as Lord Grimshaw) who did away with the fixed link between the pendulum and the pallets. In Denison's modification, as
the pendulum swung to and fro, it pushed away metal arms, each of
which was attached to a single, independently pivoted pallet. As the
pallet arm was pushed away by the pendulum rod, its other end
would disengage from one of three blades sticking out from the
driveshaft. Under pressure from the clock weight, the driveshaft
would then spin until the next blade came round and caught on the
other pallet. This was then released by the next pendulum swing,
which pushed that pallet away.

The key to the system's success was that after each pallet was
pushed away by the pendulum, as the pendulum then swung away

from it, the pallet arm would fall back by its own weight, to the stop position, where it sat, ready to halt the driveshaft blade on the next cycle.

Denison's idea made Big Ben so accurate that today, broadcast on radio and television, it has become the nation's timekeeper — so much a part of British life and culture that its chime is now as user-friendly and unintrusive as is the plastic wrap with which this historical journey began.

Clocks were one of the world's first technologies to use interchangeable parts. And the nineteenth-century American clockmakers learned their techniques from musket makers. . . .

8 Getting It Together

THE gun is often a feature in late-twentieth-century news stories seen too often around the world. In many cases the picture is of an immobile plane, full of terrified and exhausted passengers, with a hijacker holding his gun to the head of the pilot. Police sharpshooters wait in hovering helicopters, hoping for a kill, and ambulances arrive to care for the wounded. Such are the extraordinary pathways of change that, ironically, all three elements in this modern tragedy — the aircraft, the weapon and the medical technology — are linked. Each of them exists in part because the others exist.

On board the ambulance are all the modern emergency-service aids with which to treat the wounded and dying, either on release after a peaceful outcome or in the event of a violent resolution. The essential element in such emergency care today, apart from the ubiquitous antibiotics,[134] is anesthetic gas, pioneered by Paul Bert, a nineteenth-century French physiologist who spent part of his early professional life experimenting on rat-tail tissue. Bert lived in Paris and early on became known for his classic work on sensitive plants, during which time he anesthetized sensitive mimosas to see why they closed up at a touch. The movement turned out to be due to changes in the volume of the cells in tiny structures (called pulvini) at the base of each leaf blade that act to protect the plant from pressure damage.

Bert was interested in pressure of all kinds. In 1868 he built a steel chamber and started experimenting on himself, to investigate

134 152 *119*

the effects of high and low pressure on divers and mountaineers. He found that low-pressure oxygen induced high pulse rate, headache, dizziness, darkening of vision, nausea and general mental slow-down. High-pressure oxygen was simply poisonous. Bert also discovered the "bends," symptoms that occur when nitrogen bubbles form in the blood of divers who come back to the surface too quickly. But Bert's key discovery was that all gases have their effects not because of how much of a gas is present but because of the *pressure* of the gas.

Bert devoted most of his efforts trying to find out why laughing gas (nitrous oxide) could both anesthetize and asphyxiate. Finally, he tried out a mixture of ⅙ oxygen and ⅚ laughing gas at 1½ times atmospheric pressure and found that at these levels the body received enough oxygen for respiration *and* enough laughing gas for a knockout. Trials in the field led Bert to work with balloonists, who spent most of their time breathing air at different heights and pressures and in various amounts. After training a flight crew in his pressure chamber, in 1875 Bert equipped them with bags made from cow's intestines and filled with pressurized oxygen to breathe when they became dizzy at high altitude. Gaston Tissandier, the crew member who used the bag as Bert instructed, survived, but his colleague died. Bert published his research in 1878, and *La Pression barométrique* became the classic reference text for aeronautical medicine in World War I.

In Bert's time, balloons were already being flown up to twenty-five thousand feet to conduct high-altitude weather surveys, as well as for photography (the first aerial photographs of Paris were taken in 1858). Early in the nineteenth century, balloons had also been used as airborne military observation posts for the Napoleonic army.

Ballooning[135] owed its origins to papermaking, because the first balloon enthusiasts, two brothers named Joseph and Jacques Montgolfier, ran paper factories near Paris. Their profession provided them with plenty of raw material for making model balloons and for the balloon fuel. The brothers were probably inspired to their first attempts by the prize offered for an idea that would break the siege of Gibraltar. In 1781 Spanish forces had been trapped there for two years by the British, and the French were involved in the matter because they were fighting on the side of the Spanish.

The Montgolfiers thought a balloon could get into Gibraltar

135 20 25
135 69 47
135 81 53

over the heads of the British army, so they began experiments at Annonay, outside Paris, on November 15, 1782. Their first model balloon rose to a height of seventy feet, fueled by chopped hay and wool, burning beneath a taffeta silk balloon envelope. On June 5, 1783, a forty-cubic-foot balloon made of paper-covered canvas, thirty-five feet in diameter and carrying four hundred pounds of ballast, reached six thousand feet. Unfortunately, manned hot air balloon flight was not ready in time to lift the Gibraltar siege, but in 1783 the first flight took place in the Bois de Boulogne, Paris, going seven and a half miles, lasting twenty-six minutes, reaching a height of three thousand feet and carrying J. F. Pilâtre de Rozier and the Marquis d'Arlandes, two aristocratic amateur aeronauts. This followed an earlier, experimental flight a month before, when, in the presence of the king and queen, the Montgolfiers had launched a balloon crewed by a rooster, a sheep and a duck.

One of the king's other amusements, besides staring at airborne animals, was watching the elaborate Versailles palace fountains. At enormous cost these were supplied with high-pressure water from the nearby River Seine, where there was a giant, watermill-powered pumping station, expressly built for the job. Then, in 1795, when Joseph Montgolfier was on holiday at the beach, he had an idea for supplying water under pressure for much less money and involving no moving parts. He based the idea on the way he had seen the tide rushing up, explosively, through holes in the rocks. He called the machine using this concept a hydraulic ram.

In 1805 the first prototype ram was laid in the bed of a flowing river, where water under pressure flowed into a box fitted with a stop valve. At a certain water pressure the valve was forced closed, causing the water in the box to back up and enter another chamber filled with air, which the water compressed. At maximum compression the entry valve to the compression chamber closed, an escape valve opened and water held in the compression chamber was forced up a delivery pipe. The air pressure then fell, allowing more water in, opening the primary water inlet valve and beginning the cycle again. The ram could do this trick 120 times a minute.

By the early nineteenth century, seven hundred rams were in service all over France — in urban water supplies, irrigation systems and canals. James Watt's factory in Birmingham[136] bought one, and later on hydraulic rams would be used in building the Brittannia Bridge in Wales and the Hudson River tunnel in New York. But

136 17 24
136 221 184

back in Versailles, in spite of the ram's success and the government's promises, no funds were ever forthcoming to replace the *machine de Marly.*

The ram was to have political ramifications, too — in the matter of the brief unification of Italy — about fifty years after it was invented. In the mid–nineteenth century the king of Sardinia (who happened also to be king of large bits of northern Italy) decided it was time to do something about his province of Savoy, stuck on the wrong (i.e., northern) side of the Alps. Also, apart from this minor political problem, the Alps were causing other difficulties.

The entire northern European railway system was forced to stop there. Commerce was obliged, expensively, to go round the mountains by ship. So were travelers, especially those returning from the various empires in the Near and Far East. So the obstacle presented by the Alps was costing everybody a lot of money, which is always a great stimulant to innovate.

On August 15, 1857, in the presence of King Emmanuel and Prince Napoleon, work began on a tunnel (one half paid for by Sardinia, the other half by France) under Mont Cenis,[137] running from Modane, in Savoy, to Bardonecchia, near the Italian city of Turin. The engineers started by hand-drilling the holes for explosives, which meant advancing the tunnel about nine inches a day. Total time to completion at this rate was reckoned to be over forty years. So in 1861 the chief engineer, Germain Sommellier, decided to speed things up a bit, with a variant on Montgolfier's hydraulic ram that would compress air to drive pneumatic rock drills.

The ram was powered by a series of pipes bringing water from a reservoir 164 feet farther up the mountain. The falling water turned wheels that moved pistons to force water up pipes, compressing air. The usual system of valves opened and shut to keep air and water apart. The compressed air was then fed down tubes into the tunnel, to supply nine rail-mounted, twelve-ton, rotating, automatic-feed drills. At each stage of the work, the drills were moved forward and bored a total of eighty holes of various depths. All but the three center holes were then filled with gunpowder and exploded.

Thanks to the new drills, tunneling speed was improved by twenty times, to a rate of nearly fifteen feet a day. Completing the Mont Cenis tunnel (and any other like it) was not going to take a lifetime, after all. Breakthrough came on Christmas Day 1870, eight

137 179 *140*

miles, twenty-eight deaths, 2,954,000 rounds of explosives and
£3,000,000 later. Alas, Italian unity would have to wait. During the
drilling, the Italian War of Liberation had taken place, and Savoy
had been ceded to France. Still, the French had paid their half of the
costs. After the indubitable success of Mont Cenis, there was no
holding back eager investors, particularly the Swiss. Within thirty
years the St. Gotthard, the Arlberg and the Simplon tunnels were
all open, and the new Orient Express was carrying travelers all the
way from Calais to Istanbul.

By the strangest irony, Sommellier and the other three Mont
Cenis chief engineers all died of heart attacks. The irony was that
all three men had, in the course of tunneling, made extensive use of
a new explosive. Nitroglycerine had been discovered by an Italian
called Ascanio Sobrero in 1846, but its manufacture was hazardous,
to say the least. But in 1862 a Swede and his son found a way to
make nitroglycerine a little more safely. Unfortunately, the process
was not yet safe enough. In 1864 their entire factory at Heleneborg,
Sweden, blew up, killing the Swede's other son. But the search for
safety went on, and in 1867 Alfred Nobel[138] (the surviving son) was
ready to patent a mixture of nitroglycerine and a type of clay called
kieselguhr that produced a new explosive called dynamite.[139] First
used in the Mont Cenis tunnel, dynamite offered much more bang
for the money.

The irony was that the heart attacks suffered by all three Mont
Cenis project chief-engineers might well have been prevented by
the same chemical that had so successfully helped them blast out
their tunnel. Since 1867 another use for nitroglycerine was as a
pharmaceutical called glonoin (which included 1 percent nitroglycerine and 90 percent alcohol), prescribed as a vasodilator to relieve
cardiac pain and the symptoms of angina. Small amounts of nitroglycerine relax the coronary and other blood vessels, allowing the
blood flow to increase and blood pressure to fall.

One unfortunate side effect of the other, explosive form of nitroglycerine was that prolonged handling of the material caused what
was understatedly known by doctors at the time as a "dynamite
headache." This was sometimes relieved by taking nitroglycerine in
its glonoin form, to dilate the arteries at the base of the neck whose
constriction, thanks to tension (or a hangover), may cause a
headache.

138 47 39

139 48 39

This practice continued until a new medication changed the world of pain in general. In 1853 a French chemist, Charles Gerhardt, had produced a form of salicylic acid, called acetylsalicylic acid (ASA). Gerhardt's work was based on that of a German who'd originally extracted salicylic acid from the meadowsweet plant. This would cure headaches, but the extraction process was so time-consuming that he gave up. By the 1890s every German chemist was attempting to derive chemicals from coal tar,[140] the new wonder-gunk throwaway by-product of gaslight. August Hofmann,[141] working for the Bayer[142] company, found a chemical called phenol, from which *artificial* acetylsalicylic acid could be made simply, cheaply and quickly. The product was given an acronymic name made up of the letters *A* (from acetyl), *SPIR* (from *Spiraea ulmaria*, the name of the meadowsweet) and *IN* (origin unknown). These letters formed the trade name *aspirin*. Headaches would never be dynamite again.

The other name for phenol was carbolic acid. So it might be expected, with hindsight, that it would have been used for disinfectant purposes. However, it was the particular way the disinfectant ended up being dispensed that would eventually bring this historical tale back to the hijack event with which this chapter begins. In 1834 yet another German chemist, called Runge, had found that the crude form of carbolic acid (known as creosote) was good for stopping wood-rot. Then in 1857 carbolic was used in Carlisle, England, in a vain attempt to cure cattle of anthrax.[143]

By 1867 the professor of surgery at the University of Glasgow, Joseph Lister, who had heard about the Carlisle experiments, developed a dressing using carbolic acid dissolved in paraffin and backed with a layer of muslin, to create an artificial scab on a wound. Later he would use thin rubberized cloth[144] as a dressing, utilizing material originally made by the Scots inventor Macintosh for his raincoats. At the time, deaths from postoperative infections were common, though nobody knew why. However, when Lister tried the new dressing on thirteen of his compound fracture cases, all thirteen miraculously survived.

During this period, surgery often led to complications devoutly to be avoided, giving rise to the phrase "a successful operation, but the patient died." Unknown to medicine at the time, hospital conditions (unspeakable filth) were responsible for infection on a grand scale; so people generally went to the hospital expecting to die. But with Lister's next use of carbolic acid, surgery was about to become something better than butchery. Lister's success was achieved in large part thanks to the earlier efforts of a fanatically boring, abstemious bicycling fan called Benjamin Richardson who had recently made the astonishing observation that extreme cold causes numbness.

Richardson was an anesthetist; like everyone else, including Lister, he played around with carbolic acid. He found that frozen carbolic acid indeed numbs the skin, but in doing so, it also destroys the tissue. Then one night in 1867, at a ball a young woman dropped or blew some eau de cologne onto his forehead, and he noticed the cooling effect it had as it evaporated. So he went home and invented the ether spray. Ether was a known anesthetic, and now it could be sprayed locally on gums, tumors, the breast or anywhere a local surface anesthetic might be needed by the surgeon.

The link with Lister is anybody's guess, but together with a fashionable new "perfume vaporizer" being used around London parties and in the theaters at the time, sprays may have been on the medical mind. Not surprisingly, in 1871, when Lister began filling his operating theaters with a fine carbolic mist pumped by a device thought to be developed from Richardson's original rubber-ball spray bottle, Lister added a new phrase to the ever-hopeful surgeons' preoperative vocabulary: "Let us spray." The mist was effective but eventually caused complaints from patients, including Queen Victoria, because it produced a painful stinging sensation on exposed skin. So by 1887 the spray was abandoned in favor of more generally hygienic behavior on the part of the medical staff.

It was Wilhelm Maybach, a German engineer, who took things to the next stage. In 1893 Maybach was working with a better-known colleague, Gottlieb Daimler, whose sales chief had a daughter with the best-known name of all: Mercedes. Maybach used the spray concept to invent the carburetor, installing it in the new car which Daimler named after the girl. A carburetor operates by introducing a fine spray of air and fuel into the cylinder, where a spark plug explodes the mixture, driving the cylinder piston up and down. Maybach's carburetor succeeded because he used a "float feed" principle. Gasoline was fed by pressure or gravity into a small chamber that contained a float. From there, the fuel went along a pipe ending in a fine hole placed at a suitable point on the engine air-inlet pipe. Suction caused by the upward stroke of the cylinder piston caused gasoline to leave the fuel pipe and mix, as a fine spray, with air coming into the cylinder. The float controlled a needle valve regulating the amount of fuel coming from the gasoline tank into the float chamber. In this manner, the gasoline level in the chamber could be maintained with extreme precision. This float principle was adopted in almost all automobiles because it solved the problem of mixing air and fuel in the correct proportions.

The up-and-down motion of the cylinders in Maybach's engine was then converted to the round-and-round motion of a shaft, linked to cogs to turn the car wheels or anything else that needed to go round at very high speed. The process was like a turbine fan on the front end of a jet engine that spins to expel air (superheated by a burning mist of fuel), which then expands out of the back of the engine to drive an airliner such as the hijacked one with which this chapter begins.

The frightened passengers peering out of the windows of the hijacked airplane are generally in their seats for one reason above all. At some time earlier, they or their travel agent bought tickets for this fateful flight, using one of the many worldwide airline reservation systems. The first nationwide airline reservation system came into existence because of the shockingly unexpected detonation of a Russian atomic bomb, in August 1949.

The panicked American response to the explosion was a total rethink of their national defenses. The Americans introduced a massive bomber-building program and designed the first computer-linked chain of defense radars. On a three-thousand-mile arc along the northern coastline of Canada and Alaska, from Point Barrow in the east to Baffin Island in the west, more than fifty radar stations were set up as part of the new distant early-warning system (nicknamed "DEWline"). The stations were grouped in six sectors, each covering five hundred miles of coast. Signals from these far-northern stations, each of which could detect and track aircraft as far as two hundred miles away, were fed to the new National Defense Command Center under Cheyenne Mountain in Colorado. Now America had up to four hours' warning of an over-the-pole bomber attack. The computers linked all DEWline stations to the command center, synthesized and correlated the signals and presented an overall picture of defense airspace.

One day in the spring of 1953, an IBM engineer who had been working on the DEWline project happened to be on an American Airlines flight from San Francisco to New York and sat next to the airline's president. Upon discovering that they were both named Smith, they began to talk. Of such minutiae is history made. The IBM man explained the principle behind DEWline to his fellow passenger, who saw immediately that the information networking made possible by IBM's air-defense computers would make the perfect airline-reservations system. When American Airlines inaugurated the system in 1962, it was called semi-automatic business environment research, or SABER for short. It was the prototype for every reservations system that followed.

SABER linked thousands of agencies, reservations terminals and ticket desks. It also continually updated information regarding passengers' contact numbers, meal requirements and hotel and car reservations. Significantly, it was soon extended to handle flight planning, maintenance reporting, crew scheduling, fuel management and all the real-time aspects of running an airline, in a more complex manner than had ever been possible before.

SABER drove air transportation to a new level of sophistication and scale, triggering in turn Project Beacon, a new automated air traffic control system based on the same computerized principle, which entered service for the first time at the Atlanta airport in 1966. When SABER linked thousands of terminals and processed massive amounts of data in seconds, it had made possible the immense complexity of modern air travel. It also showed the business world the potential benefits such a system could bring to any industry where distributed operations needed better and more efficient organization.

In one of those extraordinary connections that history sometimes makes, the reason why SABER, and indeed computers in general, had come into existence in the first place can all be traced back to a single event. In 1798 one of Napoleon's officers, up the Nile with the Egyptian campaign, came across fine silk shawls, imported by the Egyptians from Kashmir. When he and his fellow soldiers started sending the shawls home as presents, French women went crazy for them. Napoleon's wife, the empress Josephine, bought four thousand. A pair of shawls (traditionally two were made at the same time) could take several years to complete, and their texture was so fine that they would pass through a wedding ring when they were presented as traditional wedding presents. In Kashmir the shawls had originally been the gift of princes, in return for the offerings of their vassals (*shawl*, after all, means "gift").

Once the fashion was established in Paris, it was imitated in Britain. First, the shawls were made in silk and wool for the better-off customers, and then, as the craze spread, the manufacturers produced versions in cotton for the mass market. "Cashmere" shawl factories were set up in Rheims and Lyons in France, as well as in Norwich, Huddersfield and Bradford in England and, most successfully of all, in the Scottish town of Paisley. If you own a paisley-pattern tie or scarf, this was where it started. By the late 1860s everybody was wearing cashmere patterned shawls, which became the universal wedding present. The pattern was a difficult one to copy, as it featured complex, traditional designs of pine or fir cones, representing fertility and prosperity in the Muslim world (though this fact was not publicized in stuffy Victorian England or in even more proper America, where the fashion was at its height in the last years of the century).

Herman Hollerith, a young engineer working on the 1890 U.S. census, became desperate for an automated way to count people. And now, once again, we come across one of those crossroads on the web of change that had such fundamental effect that the pathways of history cross it again and again. Hollerith's brother-in-law was in textiles and told him about a new loom that could weave patterns as complicated as those of the most expensive silk cashmere shawls without making errors. The loom (known as a Jacquard) used a control mechanism consisting of a sheet of paper with a pattern of holes in it. When sprung wire hooks were pressed against the paper, a hook would go through where there was a hole and pick up the required thread, automating the business of weaving the most complex of designs.

Hollerith adapted the idea, using paper cards the size of a dollar bill because there were dollar-bill-sorting machines already available for use. The holes in a card represented data; so, for instance, if you were a male, Greek-born carpenter living in Philadelphia, there would be a hole in the card to represent each of these facts. Sprung, electrified wires passing through a hole made contact on the other side, and the signal moved the clock hand of a counter registering the relevant bit of data. With a sorting system based on the dollar-bill holder, all the data could later be collated and analyzed at high speed. Hollerith's idea for counting, collating and analyzing figures was so successful that he went into business with some people who later changed the company's name to International Business Machines.

In 1908 the engineer getting ready for the 1910 census, John Powers (who had worked with Hollerith), was asked by the authorities to design a system sufficiently different to avoid conflict with the Hollerith patent, because the Census Bureau felt that on this occasion Hollerith's price was too high. Powers used the card concept for a slightly different purpose. He designed a machine on which the holes in the card represented figures like those on bills or in inventory data, balance sheet totals, sales figures, etc. Each card had forty-five columns and up to nine holes per column. During processing, an electric sorting machine found the cards required for collation or analysis, using a metal brush that sensed an electric current coming through when a card with a hole in it passed between it and a source of electric current behind the card.

This machine sorted fifteen thousand cards an hour and could be linked to a tabulator that would add up how many of a particular hole (i.e., a datum) had been identified, so it was ideal for statistics and records and stock control of all kinds. Bookkeeping was no longer going to need books. In 1913 this was especially good news to the collectors of the new federal income tax, who were naturally keen to automate.

But the real key to the success of the tabulator, its link to the hijack event with which this chapter begins, was the manner in which Powers's holes in the card were made. Powers designed a small keyboard, with ten keys, to punch the numbers 0–9. Each time a hole in a column was punched, the card automatically moved along, so as to place the next column under the punch. Given the design of his machine, it was logical that Powers should take advantage of the tremendous growth in business administration going on in the 1920s (and particularly in accounting, finance, marketing and management) to merge with an office-equipment company that made typewriters. The keyboard on his punch used typewriter keys, so in 1927 he became part of the Remington Typewriter[145] Company, which was itself part of Rand, the biggest office-equipment company in America. Rand made over four thousand office prod-

145 21 25

ucts, from card indexes to filing cabinets, mainly for banks, insurance companies, libraries and government agencies — all of which were interested in accurate record keeping.

All this office work had been triggered in the first place by the typewriter. In 1867 a printer named Christopher Sholes used the idea of piano keys to invent the new machine. He also came up with the modern QWERTY keyboard layout, on the principle that if the most-often-used keys were placed close to one another, they would stick. But the major problem in manufacturing the typewriter was that Sholes did not have the equipment to engineer the parts with the necessary level of precision. So in 1873 he took his design to a company called E. Remington & Sons, of Ilion, New York, and by 1888 its speed-typing contests proved such a good publicity gimmick that Remington couldn't keep up with demand. More important, it opened the world of business to women for the first time.[146]

146   76   *50*

Remington had bought Sholes's idea because, with the Civil War now over, the company was looking around for something else to do with the precision machine tools in its factories. These tools had originally been developed to make interchangeable parts and, as such, were perfect for all the bits that went into a typewriter. But back before 1873 the tools had been used for something rather different. Remington had used them to produce a new weapon; from field trials conducted with the U.S. Ordnance Department, the company's split-breech rolling block became the essential element in the most popular military rifle in the world. The breechblock and hammer were hinged on center pins, set at right angles to the barrel axis. This alignment gave a perfect seal at the moment of firing, when the two parts slammed against each other. The bullet used was of French rimfire design, and the breechblock had a slot which exposed the upper edge of the cartridge rim to the hammer. Remington sold over a million of the rifles to armies in Denmark, Sweden, Spain, Egypt, France and the United States.

It was the rimfire bullet that also made possible the development of the automatic pistol. The king of pistol design was, of course, Sam Colt, whose company was the only other to compete with Remington during the Civil War. Colt was said to have stolen the concept of his famous revolver from an Englishman while on a visit to English troops in Calcutta in 1830. Be that as it may, when he returned in 1831 with drawings and a wooden model, he couldn't afford to finance production. So he spent some time raising money as a lecturer on the medicinal properties of a newly discovered anesthetic called nitrous oxide (also known, from the effect it has, as laughing gas). To bring in the crowds, he billed himself as "Dr. Coult" and claimed scientific qualifications (all of which were bogus).

By 1836 Colt had just enough money to go into the revolver-making business, until he went bankrupt and had to think of another military money-maker. In 1841 he hit on the idea of underwater mines, forming the Submarine Battery Company and receiving $6,000 from the U.S. government to set up a demonstration. He also wangled an invitation to go to Russia for the same purpose, since at the time the Russian authorities were the only ones with an organized and systematic mine-development program. Meanwhile, by 1842 Colt finally had enough gunpowder for his American demonstration. Using an electric signal sent through several hundred feet of copper wire (which he had borrowed from his neighbor in Washington Square, Samuel Morse[147]), he blew up a box of gunpowder attached to the hull of a hulk floating in New York Harbor. Later, on the Potomac River, in front of eight thousand spectators, he blew up a moving ship, at a distance of five miles. But since Colt would not reveal how he had done it (there were rumors of towers and observers and mirrors), the U.S. navy said it could see no use for the idea. And that was the end of the matter.

In any case, Colt had already been upstaged in Russia by Alfred Nobel,[148] who showed he could do the same thing without the need for a wire or anybody up a tower to send signals. With Russian financial backing, and for reasons of military security, he set up a company with the delightful cover name of Colonel Ogarev and Mr. Nobel's Chartered Mechanical and Pig Iron Factory. Here he made mines that could be individually moored by underwater ropes to float a few feet below the surface. Each mine consisted of a two-foot container, filled with gunpowder and topped with a lead-and-glass container about the size of a pencil. This held sulfuric acid and was suspended above a mixture of chlorate of potassium and sugar. When the lead was bent by contact with a ship's hull, the glass broke and the acid fell into the mixture, triggering instant heat and flame, and causing the gunpowder to explode.

147 30 *29*
147 215 *179*

148 47 *39*

The Russians loved the technique and asked Nobel (by now, thanks to his earlier invention of dynamite,[149] acknowledged to be the world's expert on explosives) to manufacture hundreds of these mines. After first successfully protecting their Baltic harbors with the new weapons, the Russians sowed them in the waters off Sevastopol during the Crimean War. This action obliged the Allied armies to mount a difficult land attack on Sevastopol, backed up by a supply fleet which (for the same reason) had to sit at anchor off the nearby port of Balaklava. This was where the great hurricane of November 14, 1854,[150] destroyed the entire fleet, forcing the Allies into a winter campaign for which they were now woefully ill equipped.

The winter battles in the Crimea in 1854 revealed the appalling state of the British army. Reports to the London *Times* by Howard Russell about the treatment of the sick and wounded caused outrage in England. In October Florence Nightingale and twenty-eight nurses set out for the Crimea. Miss Nightingale was a formidable woman, so obsessed by the idea of nursing that she had refused offers of marriage because it would get in the way. In 1854 she was nursing cholera victims at the Middlesex Hospital in London when she read a newspaper report complaining that there were no British nurses in the Crimea. Within a few days she had taken ship.

When Nightingale arrived, what she saw beggared description. The hospital was a charnel house. There was no hospital furniture, no nurses, no dressers; the mattresses and floors were covered with filth and excrement. There were no tables on which to operate, the ratio of surgeons to wounded was 2:500, a thousand men were suffering from diarrhea but there were only twenty chamber pots, and there were neither blankets nor shirts. Chloroform and ether had been in general use in Europe and America since 1847, but nobody had supplied them for use in the Crimea — as army administrators thought that "they took too much time to administer." The wounded either died where they fell or were dragged to treatment by their comrades. Doctors on the battlefield had not been provided with any means of transportation, so by the time they arrived, through waist-deep mud and snow, it was usually too late. Some casualties made it to the hospital ship, where conditions were even worse than in the hospital. London-based military medical experts recommended smoking to kill germs, and a moustache to filter out diseases.

Nightingale was financed by £30,000 of public money raised by the London *Times*. She spent it rapidly on bare necessities like shirts, flannel, kettles, pans, socks, operating tables, combs, towels, soap, scissors and food for proper nutrition. She even persuaded the chef of the Reform Club to come to the Crimea and take charge of the hospital kitchens. Six months after her arrival, the mortality rate had dropped from 44 percent to 2 percent.

This extraordinary woman also unceasingly pressured the military authorities into setting up a sanitary commission to clean and limewash hospital walls and to remove the sewer above which the wards were situated. The following June, thanks to Nightingale's relentless lobbying, the Army Medical Corps was established.

Nightingale's one-thousand-page report on disease, and the military response to it, was produced after the war was over. It showed that although most men in the army were from the fittest sector of the male population, between ages seventeen and thirty-five, even *in Britain* their mortality rate from disease was double that of civilians. Of the 18,058 soldiers who had been killed in the Crimea, only 1,761 died because of enemy action. The other 16,297 deaths had been the result of hospitalization. When the Royal Commission on the War read these statistics, there was a storm of protest that brought down the government.

One of the people most profoundly influenced by Nightingale and her report was Jean-Henri Dunant, a Swiss philanthropist who had gone through a similar experience in 1859 at the Battle of Solferino, during the Franco-Austrian war. In 1864, as a result of his work (and boosted by Nightingale's Europe-wide publicity), the first Geneva Convention was established for "the amelioration of the condition of wounded armies in the field." At the convention the basic rules were established for the operation of a new international organization, to be called the Red Cross. Its symbol can be seen on ambulances today, including the one at the hijack scene with which this historical journey began.

Hospitals and clinics all over the world rely overwhelmingly on antibiotics, one of the world's miracle medicines, which were developed against all odds. . . .

9 The Big Spin

I N some ways science is all about prediction, because if a scientific theory is right, it predicts accurately the behavior of some process in nature. The irony is that so much of what happens in the day-to-day events that lead to a scientific discovery is anything but predictable. The process by which discoveries are made is almost always turbulent with chance or accident, and only very rarely is a discovery the result of careful, rational planning. Science, like everything else in history, often depends on how the wheel of fortune spins.

This seems to be especially true of major scientific theories. Einstein said that the idea of relativity came to him when he had a dream about riding on a beam of light. The science of bacteriology was triggered by the accidental spillage into a tissue-culture dish of a dye which turned out to stain bacteria. Hollerith invented the punch card to be used in computers because he happened to have a brother-in-law in textiles who knew of an automatic weaving technique involving perforated paper rolls. Plastic wrap came out of an accidental explosion in an English chemical laboratory.

At the beginning of any sequence of events that would eventually lead to major innovation and change, few of the people involved would have bet on the outcome. Ironically, one of the greatest of medical discoveries beat all the odds, not least because it was the end product of an extraordinary chain of events that began at the gaming tables. In 1915 the casino in the northern French resort town of Boulogne-sur-Mer had been commandeered by the Allies as a field hospital, where a team of Army Medical Corps volunteer bacteriologists gathered together to inoculate soldiers. One of the volunteers was a young doctor, Alexander Fleming, who had already made a reputation as an expert in treating syphilis.[151]

151 194 *155*

During their time at the casino-hospital, the team began to investigate the process of infection in the wounded. It soon became clear that the major causes were the particles of the soldiers' own uniforms that had been driven into the body by bullets or shell fragments. Until then it was generally accepted that the best treatment in these cases was to administer strong antiseptics, but the team showed that antiseptics often failed to penetrate the wound cavities deeply enough — and indeed sometimes killed off the body's own defense microorganisms.

When Fleming returned to London after the war, he decided to concentrate on the problem of killing bacteria without harming tissue or damaging the immune system. Now, if Fleming had a reputation for anything other than medical brilliance (and being a good marksman), it was for keeping the most untidy laboratory in London. At Saint Mary's Hospital, Paddington, he would often finish whatever investigation he happened to be carrying out and then leave the dirty tissue-culture dishes lying around, unsterilized. After some weeks the lab would be cluttered with dozens of these unattended cultures, and sooner or later he would idly examine them for anything interesting, before finally having them washed with antiseptic.

One day, in early September 1928, he returned from holiday to the usual mess. Earlier on, he had been looking at color variations among bacteria and had cultured dozens of dishes with staphylococci taken from boils and abscesses, as well as from nose, throat and skin infections. Some of these dishes happened to be lying half-submerged in a shallow tray of Lysol antiseptic, and Fleming took the usual cursory look before discarding it. To his surprise a patch of mold had grown in the middle of one dish, and around it was an area clear of bacteria, which appeared to have been killed by the mold. This accidental discovery was to earn Alexander Fleming the Nobel Prize, because the mold turned out to be penicillin.[152]

152 134 *103*

For a gambler, the exact sequence of events required to cause the penicillin to grow and Fleming to notice would have required the longest odds ever. First of all, the window had been left open, allowing the mold spores to drift in from the street outside. Then the staphylococci in the dish had, untypically, not gone through the incubation process. From July 27 until August 6 the room temperature had remained between 16°C and 20°C (60.8°F–68°F), a range in which the mold grew best. This was followed by a warm spell exactly long enough for the staphylococci to grow. On his return from holiday Fleming had happened to pick up that particular dish, rather than any other. And entirely by chance it had not slid into the antiseptic liquid, which would have killed both mold and staphylococci.

Penicillin, the product of Fleming's untidiness, works not by killing bacteria but by producing a defect in their offspring that causes the dividing cells to become swollen, to turn transparent and eventually to burst apart. The irony is that the man who had originally discovered the process of cellular pathology had an obsession for orderliness. He was a doctor named Rudolf Virchow, and because of his relentless work ethic and intolerance of laxity, he was to become known as the "Pope of German Medicine." He once said that the key characteristic for any good researcher was "icy enthusiasm."

Virchow also espoused liberal political causes. This first came to light during the winter of 1847, when he was asked to report on a typhus epidemic which had devastated famine-ridden Upper Silesia, a province inhabited by a Polish-speaking minority. In March 1848 Virchow spent three weeks there, a period he later described as "decisive" in the development of his political and scientific convictions. His report on the epidemic, based on social, anthropological and epidemiological analyses, placed the blame squarely on the government. What was needed in Upper Silesia, said Virchow, was not only medicine but the creation of social conditions which would prevent a repeat of the disaster. These reforms included full and unlimited democracy, the admission of Polish as the official language, separation of church and state, a road-building program, fair taxes, and free and universal education.

In late March, just after his return from Silesia, came the political revolution in Germany that Virchow had been hoping for and during which he found himself fighting on the Berlin barricades. That autumn he became a member of the new Democratic Congress. A year later (at the age of twenty-eight) he moved to take up a professorship at Würzburg and begin the work for which he then became famous. Virchow had tremendous drive and energy. Besides his research, he would publish more than two thousand papers and books, travel constantly and do almost entirely without sleep.

In 1851, after two years of intense microscopic work, he published *Cellular Pathology* and changed the course of medicine. The cell, Virchow said, is the fundamental unit of life, and each cell comes from another cell, through the process of division. Disease and good health both originate with the cell. Cells are neither good nor bad, they merely carry out the role for which they are designed. So the pathology of cells is the secret to correct diagnosis. And since sick cells are the cause of disease, the secret of effective treatment lies in attacking the disease at the level of the individual organ where the diseased cells are present. No longer was the patient a mysterious "whole," but merely a body formed of dozens of different organs and systems, each operating thanks to collections of cells. By defining disease in this way — as a *local* phenomenon, rather than a generalized condition of the entire body — Virchow's new theory gave surgery a tremendous boost because the surgeon now had reasonable expectation that in excising the diseased part, the disease itself would be removed. Virchow's cell theory also set the stage for the development of chemotherapy, using drugs with affinities for special cells.

His political tendencies show in his book, where Virchow compares the body to a free state of equal individuals, a democratic cell-state: "They form a free state of individuals with equal rights, though not equal abilities, which persists because the individuals depend on each other and because there exist certain centers of organization, without the integrity of which the parts cannot obtain the necessary amount of nutritive materials." It was this unique mingling of the social and the medical that made Virchow one of the first authorities on public health, and by 1870 he was advising on the design of the Berlin sewerage system. It was Virchow who coined the phrase "prevention is better than cure."

In 1874 this paragon of rectitude was approached by someone so different from him in almost every way that it is difficult to understand how the two then became close friends. The man in question was considered by many to be a confidence trickster. He was undoubtedly one of history's greater egos, a self-made man whose business practices led his first employers, the B. H. Schroder company (for whom he was a dyestuffs sales agent in St. Petersburg, Russia), to write to him in 1846: "Our worst fears have unfortunately come true. In your correspondence you take on a tone which no businessman would use. . . . Never tell us what we should do. . . . You have an opinion of your influence and power which we by no means share."

Heinrich Schliemann was not a man easily to be made aware of his own inadequacies, as his roller-coaster life was to show. In 1850 he spent two years prospecting in California during the peak years 153 1 *10* of the gold rush,[153] as a result of which he became extremely rich. By 1853 he had made a disastrously bad marriage and was back in St. Petersburg as an independent wholesaler. By 1856, thanks to having been granted certain exclusive rights, he controlled a third of the total indigo dye imports to Russia. This made him even richer. Then in 1863 he liquidated his assets and began to travel the world, complaining constantly about discomfort, catching local diseases, oblivious to political upheaval and generally making himself difficult.

In 1868 he spent a year at the Sorbonne, in Paris, learning French, Arabic, ancient Greek and Egyptian; archeology; and comparative linguistics. There he conceived of his grand plan, which was to prove that, in the *Iliad* and the *Odyssey*, Homer had been telling the historical truth about Helen and Paris and Achilles and the entire Trojan War. Schliemann determined to find Troy, and after extensive travels in the eastern Mediterranean, using Homer's descriptions as his only guide, he identified the Trojan site as a small hill in Hissarlik, in northern Turkey.

After a brief visit to Indianapolis (to establish American citizenship), he divorced his wife and married a Greek girl because (he said later) she could read Homer. He then tried and failed to buy the site in Hissarlik, and so was obliged to request official permission to dig. By 1871 he had cut a wide trench through the mound, destroying without documentation anything that he considered did not fit his entirely mistaken premise. In 1872, during a visit to Europe, almost every eminent archeologist tried to dissuade him from continuing the devastation, but all attempts were in vain. It was at this point that he was introduced to Virchow, by the Homeric scholar and British prime minister, Gladstone. Apart from his medicine, Virchow had also written about the faces sometimes found on

Greek urns. Schliemann saw that the respectable Virchow might also help repair his tarnished image and asked the great man to return with him to "Troy." For some extraordinary reason he never explained, Virchow accepted the invitation.

In 1873 Schliemann found what he called "King Priam's Treasure," a massive collection of gold objects (which are now known to be a thousand years older than he thought), many of which then secretly found their way back to his family. When he had paid the substantial Turkish government fines for this illegal act, Schliemann moved on to Mycenae in Greece to look for the home of the Greek heroes of the Trojan War. This once again was an illicit venture, opposed in vain by the Greek authorities. In 1876 Schliemann found another giant collection of gold, ceramics and weapons. One of the objects looked like a death mask and was enough for him to send the telegram (which made him famous all over the world): "I have gazed upon the face of Agamemnon!"

All this time, Virchow acted as Schliemann's guide and mentor in things organizational and scientific, persuading him to collate and annotate, document and photograph, so as to deflect the worst of the criticism. However, by then, thanks to every claim being wilder than the one before, Schliemann was the laughingstock of the academic community. In 1888 Virchow accompanied him up the Nile on the start of another adventure, which came to nothing. Two years later Schliemann was dead.

Why had the virtuous and upright Virchow taken up with this devious character? Because since 1840 one of Virchow's great interests — one, perhaps, which allowed him to link medicine and politics — was anthropology. He had published numerous studies of skull shapes, had researched the anatomy of young gorillas and in 1886 had completed a survey of over six million German children, in which he discovered that there was no racial uniformity among them. Gladstone's original recommendation of him to Schliemann had been in regard to Virchow's ability to identify the racial characteristics of the faces on Greek urns.

Virchow's interest in faces and skulls sprang from the work of J. F. Blumenbach, professor at Göttingen University, who is best known for inventing "Blumenbach's position." This hypothesis involved identifying the general racial characteristics of a skull by placing it between the feet and looking down at it. Though the method sounds bizarre, it must be remembered that this was a time when it was believed by many scientists that there were such things as people with eyes in their shoulders, or who hopped on a single leg, or whose feet pointed backward, or whose willpower could be measured by the angle between spine and neck. Staring down at skulls was as good a method as any.

Blumenbach spent his life collecting and classifying skulls. Early on, until being persuaded otherwise, he dug them up during nightly visits to cemeteries. His aim was to classify the human race by the shape of the head; and in pursuit of this goal, Blumenbach invented the five basic racial types which, until recently, were standard reference terms: Caucasian, Mongolian, Ethiopian, American and Malay. Typically for his time, he considered the Caucasian to be the original race from which the others had degenerated. Blumenbach also included in his classification the elements of color, hair and body structure. In 1795 his book *The Human Species* laid the foundations of the science of anthropology.[154] However, Blumenbach was the first to point out that the human race was much more similar than it was different and that all humans shared the same basic characteristics. His collection of skulls became world-famous, and scientific pilgrimages were made to Göttingen to see it.

The mania for the classification of nature inherited by Blumenbach was common throughout the seventeenth and eighteenth centuries and had begun with the activities of an English botany expert called John Ray.[155] In 1662, at a time when the monarchy had been disestablished, and Oliver Cromwell[156] was running the now-republican Commonwealth of England, Ray was forced to give up his job at Trinity College, Cambridge, because his royalist sympathies were unacceptable. So he spent the rest of his life looking at plants. Over a period of forty-two years, he published five major (and massive) works on plants around Cambridge, on plants of England, on a new method of plant classification, on plants of Britain and, finally, his greatest work, the three-volume and encyclopedic *History of Plants*, which took him eighteen years to complete. Ray's single-minded devotion to his work and his extensive journeys of observation (which, at a time when a day was needed to cover twenty miles, took him all over Britain, as well as Germany, Switzerland, the Netherlands, Malta, Italy and France) made botany a science and gave modern botanists most of the taxonomic techniques they still use today.

Ray reckoned that proper classification required the study of the whole organism, so he looked at the form and growth of seeds, making the first attempt to define what it was that constituted a plant species. First, he made a list of differences in size, scent and taste. Then he included color and shape of the root, number of angles in the stalk, variety and curl of the leaf, color and germination of the flower, size and taste of the fruit, and color and shape of the seed. Ray recognized that huge variations could be caused by climate, soil, seasonal change and so on. It was Ray who divided plants into the modern categories of monocotyledons (with one seed lobe) and dicotyledons (with two).

Ray's thirty-chapter *History of Plants* included a total of over six thousand species, together with in-depth descriptions, notes on locality, details of habitat and relations to similar forms of plants. He expanded some of these notes into short essays, and in one case (in the discussion of vine types) he created a wine list. After he had finished with botany, he personally observed and described the life cycle of plants which he had grown from seeds or which had been sent to him by the small army of collectors he had persuaded to find and forward specimens.

When Ray began his work, there was no agreement on plant names, specific characteristics or classification. Many "recorded" species were actually nonexistent or confused with others, and descriptions were misleading or incomplete. When he finished, botany had a modern handbook and the scene was set for the Swedish botanist Linnaeus, who would get all the glory and whose Latin binomial classification system is still used by every botanist today. Toward the end of his life, Ray wrote volumes on zoology, compiled a collection of English words, collected butterflies and produced a history of fish.

Ray's forte was making lists, which sprang from a general concern among all his contemporaries that knowledge was proliferating faster than people could manage. The cause of this state of affairs was a discovery that had rocked epistemology to its classical foundations. It was the discovery of America. No classical or biblical author (the twin pillars of knowledge) had mentioned its existence. No sooner had the first foot been set in the New World than reports and samples of previously unknown organisms started flooding back to Europe. The situation was made worse in the early seventeenth century, when similar novelties began arriving

157 11 *18*   from the East with the Dutch,[157] Portuguese and Spanish explorers.

     Plants appeared for which there was no reference in the classical literature. These included chocolate, custard apples, cashew nuts, guavas, papayas, chilies, maple syrup, pineapples and tobacco. There were also animals which had not been on board the Ark, so where had they come from? The tropical rain forests of the New World broke Aristotle's weather law, which said that the farther south one went, the drier it became. These discoveries followed similar shock-

158 130 *97*   ing revelations from the astronomers. Galileo[158] and then others

158 247 *210* proved that the Earth was not at the center of the universe and that the heavenly bodies (the Sun and Moon, for instance) were not the perfect spheres they were supposed to be. There were spots on the Sun and mountains on the Moon.

     It was becoming clear that an entirely new way of gathering knowledge had to be found. Classical and biblical sources were untrustworthy, and in any case there was no single, clear epistemological system that would guarantee explanations that would bear critical examination. Then in 1624 an aristocratic English lawyer named Francis Bacon came up with a way to save the day. The business of knowledge would have to start afresh. To give such efforts direction, he published a work called *The New Tool* (Aristotle's method of investigation, now discredited, had first been published in a work called *The Tool*). Bacon wrote that the only way to survive the avalanche of novelty was to collect as much data as possible, personally. If not, data could be gathered from trustworthy sources. The data should then be ordered in lists, so that analysis could be attempted without confusion (this would be the system that Ray would use).

     Bacon also suggested that the use of numbers would be by far the best way to make the analysis of the lists reliable. This last point particularly appealed to those in the government bureaucracy. At a time of rising economic activity boosted by exploration, increasingly easy transportation systems across Europe, the establishment of colonies and war, it was important for national planners and tax collectors to find an accurate way to count the number of people in the population. In 1661, in England, there was already the means

159 273 *233* to do so — though nobody knew it until John Graunt[159] recognized the value of the London Mortality Bills (which had been published

regularly for seventy years) and wrote a pamphlet on the subject. The bills were lists, drawn up by local authorities, of the number of christenings, marriages and deaths in a parish. Graunt's pamphlet showed how useful these numbers could be in calculating the population, and so he was directly influential in establishing general registers throughout Europe. In this sense, he could be said to be the first statistician. Graunt's work was also instrumental in making life insurance a profitable business. In 1693 Edmund Halley, the eminent mathematician and astronomer, constructed a table, based on the mortality bills of Breslau, Germany, establishing for the first time the importance (in calculating population figures) of an individual's probable length of life at any time. In 1771 the next step was taken by Richard Price, a liberal Welsh Presbyterian minister who was interested in math and who lived in London (where his congregation included Mary Shelley's[160] mother, Mary Wollstone-craft).

160 255 *215*

From 1760 on, many new insurance companies started up, offering annuities to widows and old people. Price warned that the services of these companies were all based on plans "alike improper and insufficient" and that all they were doing was robbing the public of their hard-earned savings. In 1771 Price did something about it, when he published *Observations on Reversionary Payments*. This work included a mortality table, based on the bills published by the particularly well-administered parish register of All Souls, Northampton. Price called it the *Northampton Table*, and it was immediately employed (as was Price) by the recently founded Equitable Society insurance company.

Price's table calculated the life expectancy of a newborn baby in Northampton as 26.41 years. Multiplied by the number of births in a year, the table would provide the population figure, assuming parity of births and deaths. The table made it possible for insurance companies to calculate their premiums much more accurately, set their prices according to the client's age and attract more customers by a mathematical way of doing things. The Equitable Society's position was strengthened when it took Price's advice to insist on medical examinations for those potential clients "subject to vices or latent disorders."

Price's liberal tendencies took him to clubs frequented by other liberals, and it was at one called the Honest Whig that he met and became friends with Benjamin Franklin and Joseph Priestley. As the American Revolution drew near, Price was a staunch supporter of the Americans, urging them to fight for their liberty. When the War of Independence ended, Congress invited him to advise them on federal finances, but he modestly declined. A little later Yale conferred on him (and, at the same time, George Washington) an honorary doctorate.

His friend Priestley was rather less fortunate. Like Price and many other scientifically minded types at the time, Joseph Priestley[161] was a member of the free church. Because of an unfortunate speech impediment, he had failed as a preacher and turned instead to teaching in one of the new academies set up to educate the sons of free churchmen. Because of their opposition to the Established Church of England, members of the free churches were not permitted to attend university.[162] While Priestley was in London for a course in speech therapy, he met Price and Franklin and, through them, many of the major scientific thinkers of the day.

Priestley decided that science was good for progress, planning to write a number of books on its history. In 1767 he published his first (and only) book on the subject, *The History and Present State of Electricity, with Original Experiments*. Priestley rapidly established a name for himself as an experimenter and over the next decade made a number of basic discoveries, including that of oxygen. From his work on optics and astronomy, to his great delight he was invited to become the expedition astronomer on Captain Cook's[163] second voyage in 1771. But opposition to his liberal religious and political views cost him the job.

Nonetheless, Priestley[164] decided to make a contribution to the success of the expedition and set himself to invent a drink that would cure scurvy. During his experiments at a brewery near his home in Leeds, he had discovered the properties of the carbon dioxide (he called it "fixed air") given off by the fermenting beer vats. One of these properties was that when water was placed in a flat dish for a time above the vats, it acquired a pleasant, acidulous taste that reminded Priestley of seltzer mineral waters.

Experiments convinced him that the medicinal qualities of seltzer might be due to the air dissolved in it. Pouring water from one glass to another for three minutes in the fixed air above a beer vat achieved the same effect. By 1772 he had devised a pumping apparatus that would "impregnate" water with fixed air, and the system was set up on board Cook's *Resolution* and *Adventure* in time for the voyage. It was a great success. Meanwhile, Priestley's politics continued to dog him. His support for the French Revolution was seen as particularly traitorous, and in 1794 a mob burned down his house and laboratory. So Priestley took ship for Pennsylvania, where he settled in Northumberland, honored by his American hosts as a major scientific figure. Then one night, while dining at Yale, he met a young professor of chemistry. The result of their meeting would change the life of the twentieth-century American teenager.

It may have been because the young man at dinner that night, Benjamin Silliman, was a hypochondriac (rather than the fact that he was a chemist) that events took the course they did. Silliman imagined he suffered from lethargy, vertigo, nervous disorders and whatever else he could think of. In common with other invalids, he regularly visited health spas like Saratoga Springs, New York (at his mother's expense), and he knew that such places were only for the rich. So his meeting with Priestley moved him to decide to make the mineral-water cure available to the common people (also at his mother's expense).

In 1809 he set up in business with an apothecary named Darling, assembled apparatus to "impregnate" fifty bottles of water a day and opened two soda-water fountains in New York City, one at the Tontine Coffee House and one at the City Hotel. The decor was hugely expensive (a lot of gilt), and they only sold seventy glasses on opening day. But Darling was optimistic. A friend of Priestley's visited and declared that drinking the waters would prevent yellow fever. In spite of Silliman's hopes that the business would make him rich, by the end of the summer the endeavor was a disastrous flop. It would be many more decades before the soda fountain became a cultural icon for American youth.

Silliman cast around for some other way to make money. Two years earlier he had analyzed the contents of a meteor that had fallen on Weston, Connecticut, and this research had enhanced his scientific reputation. So he decided to offer his services to mining companies, as a geologist. His degree had been in law: he was as qualified for geology as he was to be Yale professor of chemistry. The geology venture prospered, and by 1820 Silliman was in great demand for field trips, on which he took his son, Benjamin, Jr. When he retired in 1853, his son took up where he had left off, as professor of general and applied chemistry at Yale (this time, with a degree in the subject). After writing a number of chemistry books and being elected to the National Academy of Sciences, Benjamin, Jr., took up lucrative consulting posts, as his father had done, with the Boston City Water Company and various mining enterprises.

In 1855 one of these asked him to research and report on some mineral samples from the new Pennsylvania Rock Oil Company. After several months' work Benjamin, Jr., announced that about 50 percent of the black tarlike substance could be distilled into first-rate burning oils (which would eventually be called kerosene and paraffin) and that an additional 40 percent of what was left could be distilled for other purposes, such as lubrication and gaslight. On the basis of this single report a company was launched to finance the drilling of the Drake Well[165] at Oil Creek, Pennsylvania, and in 1857 it became the first well to produce petroleum. It would be another fifty years before Silliman's reference to "other fractions" available from the oil through extra distillation would provide gasoline for the combustion[166] engine of the first automobile. Silliman's report changed the world because it made possible an entirely new form of transportation and helped turn America into an industrial superpower.

By an extraordinary twist of fate, within a few decades the oil industry Silliman started would come to rely, for its ability to find petroleum, not so much on chemists or geologists but on fossil-hunters. And by an even stranger coincidence, these investigators of the ancient and petrified owed *their* professional existence to an earlier revolution in transportation. And it, too, had helped create an industrial superpower, nearly seventy years previously: England.

At the time, the major problem inhibiting England's industrial development was the state of the roads. So the introduction of waterborne transportation on the new canals triggered massive economic expansion because these waterways transported coal (and other raw materials) much faster and cheaper than by packhorse or wagon. In 1793 a surveyor called William Smith was taking the first measurements in preparation for a canal that was to be built in the English county of Somerset, when he noticed something odd.

Certain types of rock seemed to lie in levels that reappeared, from time to time, as the rock layer dipped below the surface and then re-emerged across a stretch of countryside. During a journey to the north of England (to collect more information about canal-construction techniques), Smith saw this phenomenon happening everywhere. There were obviously regular layers of rock beneath the surface which were revealed as strata where a cliff face or a valley cut into them. In 1796 Smith discovered that the same strata always had the same fossils embedded in them. In 1815, after ten years of work, he compiled all that he had learned about stratification in the first proper colored geological map,[167] showing twenty-one sedimentary layers. Smith's map galvanized the world of fossil-hunting.

167 259 216

As ever, when a new technique appears, it is rapidly appropriated for work in areas of specialized expertise to which claims are jealously staked. In this case, following Smith's discovery, it was a French geologist, Alcide d'Orbigny, who had chosen to make his special study that of a particular kind of fossil which could only be seen through a microscope. He called the species *foraminifera*. The organisms were so small that a new name also had to be coined for their study: micropaleontology. D'Orbigny spent seven years examining these tiny single-cell marine organisms, whose size ranged from .01 to 100 mm. Living foraminiferae appear at all levels of the ocean and, as fossils, in all marine sedimentary rocks from almost every epoch.

D'Orbigny worked with the single-mindedness of all obsessives. He classified five classes, fifty-three genera and six hundred species of foraminiferae, based on the number and arrangements of the chambers of their tiny shell. Then he went on to greater depths and more detail. Between 1850 and 1852 he classified all twenty-eight known strata according to the fossil mollusks and invertebrates found in them. He further identified and described all eighteen thousand fossils involved. He noted that most fossils in one layer were not found in the next, so he set up a system known as the chronological table of Paleolithic time zones, which is still used today.

This was when fossils and oil came together. It was discovered in the late nineteenth century that certain different kinds of foraminiferae were always found in the same sequence when wells were sunk into the kind of rock formations that were going to yield petroleum. All that was needed was a method of locating the rock formation.

Such a method was provided by a Russian aristocrat called Boris Golitsyn who taught physics at Moscow University beginning in 1891. Golitsyn invented a device that would measure shock waves traveling through the Earth. In its final version, which he produced in 1906, this device consisted of a weight suspended on a vertical spring. Attached to the weight was an electrified copper coil, and coil and weight were suspended between the poles of a magnet. When the earth shook and the weight swung, the coil moved within the magnetic field, generating an electric charge that varied according to the coil's oscillations. This electric charge then caused a pen to vibrate as it moved across photographic paper revolving on a drum, producing the familiar wiggly-line trace of a seismograph. In 1923 a series of tests carried out on the Texas coastline showed that seismographs would make it possible to identify the particular signal that was produced when an explosive shock wave was reflected from the kind of rock strata that were most likely to contain petroleum.

Golitsyn's technique serves one other vital purpose in the modern world. It gives warning of impending earthquakes. Nobody knows why they happen, possibly because of volcanic activity, the movement of the Earth's tectonic plates, planetary wobble, even the gravitational influence when certain planets are in line. But it has been calculated that in any one year the probability is that there will be two quakes of a strength greater than 7.7 on the Richter scale, seventeen between 7.7 and 7.0, one hundred between 6.0 and 7.0 and fifty thousand between 3.0 and 6.0.

So thanks to French paleontologists, English surveyors, American oil, soda water, insurance, anthropologists, treasure-hunters and German pathologists, the next time an earthquake strikes, survivors can thank Boris Golitsyn's seismograph for any advance warning they receive. After it strikes, given the way the wheel of fortune spins, they may find they have to thank Fleming's penicillin.

Oil exploration triggers new technologies and systems in its constant search for what lies beneath the ground. That search began centuries ago, for different reasons, but with the same innovative results. . . .

10 Something for Nothing

THIS historical journey shows how a single discovery, over three centuries ago, was to trigger the eventual appearance of three very different, rather ordinary modern inventions: heat-resistant ceramics, the gyroscope and the fuel cell. These three inventions made possible a fourth one, which is one of the most extraordinary vehicles on the planet.

The story starts in mid-seventeenth-century Europe. At the time, a mining boom was in progress — or it would have been if the miners hadn't come across a serious problem. For some strange reason, the suction pump (the age-old system for draining the mines of the water that plagued a miner's life) wouldn't lift water more than thirty feet or so. At this point the miners had to introduce another suction pump, which would lift water only the next thirty feet, and so on. Several hundred feet below ground, this system became a serious waste of time and money. So concern was expressed to the authorities and thinkers of the day.

In 1639 the Genoese governor Giovanni Baliani told his friend the scientist Galileo Galilei[168] that he'd had the same trouble trying to siphon water along a pipe and over a hill. Galileo gave the job of finding the answer to his pupil Evangelista Torricelli, who told various Roman friends. In 1641 one of them, Gasparo Berti, rigged up a device consisting of a tube full of water, with its lower, open end immersed in a flask of water and its upper end closed with a tap. When the tap was opened, the water in the tube ran out into the flask until the level in the tube had fallen to a point about thirty feet above the water level in the flask below, whereupon it stopped.

168   129   97

Torricelli tried the same trick with mercury; since mercury is denser than water, he needed much less of it and could experiment on a more manageable scale. It behaved the same way as the water. Torricelli also told Marin Mersenne, a cleric (and the man with the Continent's biggest address book of amateur scientists), who in turn told the whole of Europe. The result was that on September 19, 1648, just outside the central French town of Clermont-Ferrand, a bunch of local worthies and the brother-in-law of famous French mathematician Blaise Pascal[169] (who was a pal of Mersenne's) climbed the nearby Puy de Dôme mountain, with a tube of mercury upended in a dish of mercury. At the bottom of the hill they repeated Torricelli and Berti's test. The mercury in the tube fell a certain distance and then stopped. As they climbed the mountain, the level of mercury in the tube gradually fell farther. And as they descended, it rose again. This clearly meant that the atmospheric air pressure, pushing down on the surface of the mercury in the dish, was supporting the column of mercury in the tube. That was why the miners' suction pumps wouldn't work vertically more than thirty feet. Above that height, air pressure was insufficient to support the column of water.

The miners' problem was solved, but another was created: as the experimenters had climbed the hill and some of the mercury had run out of the inverted tube, the falling mercury column had left a small empty space at the top of the tube. This space could not be air (since none had bubbled up through the mercury), so it had to be a vacuum.[170] According to the pope, this was impossible. A vacuum would mean totally empty space and — according to the church — God was supposed to fill the whole of space. So the existence of the vacuum was a dangerous thing to mention in a Catholic country. Which was why research from then on moved north of the Alps, into Protestant lands.

But in spite of the relative freedom of investigation there, it would take two hundred years and a major disaster in Russia for the world to reap the full benefits of the original mountain climbers' airless discovery.

It would be the first in a sequence of events triggered by the vacuum that would lead to one of the three inventions in the modern world with which this tale ends.

Since the vacuum in the top of the mercury tube went up and

169 272 *233*

170 220 *184*

down according to the altitude, obviously the air pressure on the mercury in the dish changed with height. It wasn't long before experimental amateurs in the English Royal Society noticed that this vacuum also rose and fell with changes in the weather. In consequence, by 1708 few gentlemen of quality were without the hi-tech wonder-instrument that this discovery had generated. The instrument became known as the barometer[171] (colloquially, the weather glass), and any experimenter worth his sodium chloride had to have one. By 1720 it was known that the mercury would stand thirty-one inches up the tube in very dry weather and only twenty-eight inches in storms. Ladies were advised to have a weather glass nearby to consult when they dressed, so as to "accommodate their habit to the weather." 171 269 *226*

By early in the nineteenth century, weather data had been recorded for long enough to show that, on average, it rained every other day, more frequently at night, most in autumn and least in winter, most in October and least in February, and that one year in five was very dry, and one in ten very wet. Not surprisingly, the reason for this interest in the weather was financial. Every year storms were sending thousands of pounds' worth of cargo (and, on average, two thousand crew members) to the bottom of the sea.

In 1835 a researcher at the Philadelphia Franklin Institute called James P. Espy (about whom remarkably little is known) started collecting data from up and down the east coast of America, as far north as Canada and along the coastline of the Gulf of Mexico. He sent out hundreds of circulars to sailors, to lighthouse[172] keepers and to six hundred army posts. After four years he had received enough data to draw over a thousand storm-track maps, from which he was able to deduce that storms were either round or oblong. By 1839 Espy was world famous, lecturing in the United Kingdom and Europe, known as the author of *The Philosophy of Storms*. Interest in the weather was now like the barometer in good weather. It was rising. 172 285 *249*

As early as 1843 England's Committee on Shipwrecks was recommending that every vessel carry a barometer. On land the first weather map was produced by a British navy captain Glaisher in June 1849 for the *Daily News* and sold to the public for a penny. Naturally enough, naval interest in things meteorological was high both in Europe and in the United States. In Washington, D.C.,

Matthew Maury, a young navy lieutenant, amassed a mountain of data on winds and currents copied from captains' logs over a period of nine years. In the end he had the equivalent of a million days' worth of observations. This was enough to persuade everybody to get together at the world's first weather conference in Brussels in 1853. As a result of this conference, nothing happened.

173 150 *116*   Then, a year later, on November 14, 1854,[173] the Crimean War Allied fleet found itself anchored off the Russian port of Balaklava. The fleet consisted primarily of supply ships like the HMS *Prince*, the greatest steamship of the time, but there were warships, including the *Henri IV*, pride of the French navy. During the afternoon, violent rainstorms began to lash Balaklava, and by nightfall a full-scale hurricane was blowing. By morning the entire fleet had been sunk. The *Prince* alone took seven thousand tons of medical supplies, boots and forty thousand winter greatcoats to the bottom. As a result of the losses, Allied troops ashore then had to live through a terrible winter without supplies. The appalling conditions under which they survived and the consequential high death rate hit the headlines everywhere, triggering, among other things, Florence

174 101 70   Nightingale's[174] famous mission, which would lead to the total reorganization of military, and then civilian, hospitals all over the world.

Within days of the storm, the French emperor Napoleon had ordered an investigation. His academician Urbain Leverrier collected 250 sets of weather data from every European observatory for the period November 11 to 16. His report, issued at the end of January 1859, shocked everybody. The Crimean fleet could have been saved because it could have had at least twenty-four hours' warning. The day after receiving Leverrier's report, Napoleon ordered a telegraphic storm-warming system to be set up all over France. In the United Kingdom the response was just as swift. There the board of trade established the Meteorological Office, and by the summer of 1860 it was receiving daily data from fifteen stations around the British Isles, as well as exchanging telegraphed weather reports, via Paris, with a dozen places throughout Europe.

Interest centered on the weather in and around Oxford Street and the London Embankment. This was where major new roadworks had been under construction through the second quarter of the century. By 1850 London had a rapidly rising population, and evidence of its mobility around the city manifested itself in the form of the twenty-four tons of manure being collected daily from Oxford and Regent Streets alone. With true Victorian mania for statistics, the planners reckoned this amount represented eight thousand daily horse droppings. In addition to the mud and loose stones with which the streets had previously been paved, cleanup costs and inconvenience were considerable. A gold medal worth thirty guineas was offered for a new street-cleaning technique. In 1844 the *Times* of London reported on progress so far: "Forty to fifty men and boys . . . employed during the day in the removal of excrementitious matter . . . [proves] that the experiment of washing the streets . . . has been . . . found successful."

In 1847 the local commissioners for roads began to introduce a new surfacing technique, invented by John McAdam, which consisted of firm roadbed foundations topped by granite chips or paving stones. At last, the streets could be washed down easily. But beginning in 1855, this would cause nothing but complications for the engineers digging London's massive new sewerage systems.[175]   175 *23 26*
Road-washing and stormwater runoff from the new impermeable road surface represented a major factor in estimating how big sewers ought to be. This calculation was accomplished with the aid of meteorological data and by referring to information collected in Paris and Lyons, where the French already had rain gauges set on nearby watersheds to warn of impending flooding.

Water and sewage discharge was analyzed as carefully as is electric power use today. In the morning, water would be emptied from baths. Before noon, kitchen washbasins, water closets and housemaids' sinks would be "freely discharging various foul matters." In the afternoon and evening, kitchen sinks would contribute most. The average sewage flow for London was reckoned to be 15,208,083 cubic feet per day. There was an average 8 percent fluctuation in peak hourly flow.

These statistics were made more complicated by the weather. Rainfall records indicated twenty-one days a year when more than a quarter inch fell. Heavy rains could cause disaster in many old sewers if the average sewage flow were less than 2½ feet per second for more than a few days. And severe storm flow could be as much as six times that of the flow in dry weather. Pipe leakage was an additional major problem. Nonetheless, the engineers pressed ahead. When the new sewers were finished in 1871, they stretched for eighty-three miles beneath the streets of London to the suburbs in the east (downriver, where the sewage would be dumped), and the system drained over a hundred square miles. The construction had cost the staggering sum of £4.6 million sterling.

Now, people spend that kind of money only when they're desperate. The cause of the desperation was an epidemic that visited Europe three times between 1831 and 1866, and in England alone killed over a hundred thousand people. It was cholera,[176] and it created total panic because it struck both high and low throughout the land. Nobody was safe, and nobody knew how to stop it. All they knew was that it hit the poor and dirty harder than the rich and clean. And there were more poor and dirty than ever before. In the overcrowded and exploding industrial cities, immigrant factory labor from the country villages packed into tenements, living in unspeakable conditions of filth, starvation and disease. One in two children died before the age of five, and only one in six adults made it to the age of fifty. Between 1831 and 1841 the average death rate from starvation and disease rose by a third. The country was already close to anarchy when cholera struck.

The Victorian middle class knew it was on the verge of revolution unless something drastic were done. So they built sewers, started Associations for the Diffusion of Sanitary Knowledge and sent their children out of the cities to schools where sports and muscular Christianity and cold showers might protect against the epidemic. They fumigated and whitewashed the slums, and Methodists held anticholera pray-ins. The marriage rate rose. Local authorities offered anticholera candies to postmen, and national compulsory vaccination was introduced. By 1875 Parliament had enacted no fewer than twenty-nine pieces of sanitary legislation, ranging from the Public Health Act to nuisance removal measures. Most important of all, the government removed the excise tax on soap; so between 1841 and 1861, its use doubled. Cleanliness was now next to godliness. Some 140,000 tracts were distributed on the "power of soap and water."

All this was music to the ears of pottery makers, who saw the greatest-ever new market opportunity emerging from the filth, death and degradation. Potters who had spent generations making earthenware Toby jugs now made a fortune glazing sewer pipes. The Doulton firm, among others, became a household name with

176 194 155

its new ceramic kitchen sinks and pitcher-and-bowl china toilet sets. Plumbing soon linked houses to the new sewers, and then everybody wanted a water closet. Being Victorians, they wanted them respectable, so the manufacturers made them in decorous white vitreous china, painted with designs called Magnolia, Wild Rose, Unitas, Renaissance, Baronial, Olive Green Chicago and Morning Glory. For those more concerned with their sanitary efficacy, names included Directo, Precipitas, Inodoro, Rapide and Deluge. The first properly modern flush lavatory was sold in 1884 by George Jennings (it won prizes and introduced the oval seat design), under the brand name of Pedestal Vase.

The first of this chapter's three modern inventions finally emerged when Victorian hypochondria brought new Industrial Revolution techniques in cast iron together with the kind of specialist glassworking skills which the new chemical industries were demanding, especially in Germany. Chemical producers needed containers that would be impervious to the effects of corrosive liquids. For this reason, the Germans (the Continent's best chemists) were particularly good at enameling. Since the 1840s they had been treating cast-iron cooking pots, and then all kinds of kitchenware, with the blue[177] enamel which would become so familiar to Europeans and to American frontier settlers by the end of the century.

177 241 205

The new boom in sanitary ware boosted the ceramics industry as never before. By 1904 no fewer than seventy German companies were producing cast-iron baths covered with vitreous enamel which was little different from the kind found in bathrooms today. And the process of manufacture had been refined to a point where the most complex of shapes could be covered using liquid enamel (made by producing the ceramic material in high-temperature furnaces, then running the molten mixture in cold water to splinter it and grinding the splinters down to fine dust). At this point, the dust was remelted into liquid at 1,000°F (542°C), and the hot cast-iron ware was either sprayed with it or dipped into it. A final dusting with glass powder gave the necessary smooth finish.

This ceramic technology was the first of the three modern inventions which come together at the end of this chapter and make possible one of the most extraordinary feats of exploration in history.

The second of these three modern inventions came about as a

result of the rapid seventeenth-century development of the vacuum as an experimental tool. Creating a vacuum involved sucking air out of the instrument which was to contain it. The first step toward being able to do that was taken in 1654 by Otto von Guericke, mayor of the German city of Magdeburg, when he designed an air pump. In London a Royal Society experimenter, Robert Boyle, used a modified version of the pump that would either increase or decrease the air pressure in a vessel. Robert Boyle placed a barometer in such a vessel and saw that as air pressure rose, the mercury in a barometer caused trapped air to become squashed. When pressure was released, the trapped air would expand and push back on the mercury. Air, apparently, was flexible, and could be compressed.

Various schemes were suggested to utilize this exciting new property. In 1653 the Frenchman Denis Papin designed the first pneumatic tube for delivering parcels. In 1785 a Scotsman suggested raising sunken ships with compressed air balloons, and after 178 150 *116* the great storm of 1854[178] some of the ships sunk in Balaklava were raised this way. The first professional diving also began.

It was news about one particular use of compressed air that would help open up the American West. In 1857 work had begun 179 137 *106* on the first tunnel under the French Alps,[179] which at the time were Italian, as Savoy still belonged to the king of Sardinia. The plan was to use the tunnel to unite the railway systems of Italy and northern Europe. This would cut the journey time between Paris and Turin to eighteen hours, radically improve freight delivery between the Mediterranean and the north, open up the Italian Riviera to tourism and shorten the British return journey from India by several weeks.

Elsewhere, this first Alpine tunneling work (the tunnel under Mont Cenis, running from Mondane in France to Bardonecchia in Italy) raised even more ambitious hopes. One day in Pittsburgh a young American bought a magazine subscription because he liked the looks of the girl selling them. The first copy he read carried a story about Mont Cenis and the compressed-air rock drills being used there for the first time (and which were increasing tunneling progress, over hand-drilling techniques, twentyfold).

The young magazine reader was George Westinghouse, and like 180 26 *28* many Americans he was well aware that railroad[180] travel was a 180 74 *50* risky business. Every month another horror story hit the headlines, as overloaded freight or passenger trains with inadequate brakes crashed on steep gradients, with horrifying loss of life. At the time, every car carried a brakeman, and in an emergency, his operation of the manual braking system was supposed to coincide with those being applied on all the other cars. Casey Jones was the most famous example of what could happen if things went wrong, and cars jackknifed into each other with disastrous consequences. As one contemporary columnist wrote: "Another railway accident (so-called) on the Erie Road. Scores of people smashed, burned to death or maimed for life. We shall never travel safely till some pious, wealthy and much-beloved railway director has been hanged for murder."

Westinghouse realized that the answer lay in compressed air. By 1869 he had a patent for the Westinghouse air brake, which has been used in some form ever since. In its automatic version, compressed air in a pipe running the length of the train (and backed up by compressed air reservoirs under each car) held back a piston. If the air pressure were released (deliberately or by accident), the piston closed, putting on the brakes. The system was a (non-) smash hit. By 1876, 15,569 locomotives and 14,055 cars were equipped, and the system could stop a 30-mph, 103-ton train in 500 feet, three times shorter than the old way.

Through his invention, Westinghouse became involved with a Croat genius who wore a new red-and-black tie each week, used handkerchiefs once and threw them away, hated billiard balls and lived in a hotel room full of pigeons. (But more about him later.) The Westinghouse brake and the now-minimal risk of crashing meant trains could run faster and more frequently. It was important to be able to control traffic movement better, so in 1880 Westinghouse began buying up patents in signaling systems. By 1882 he had set up the Union Signal and Switching Company in Pittsburgh.[181] At first he tried controlling the signals along the track by compressed air, but the pipes froze in winter. Electricity was clearly going to be the answer, and his idea was to send an electric current that would operate a valve to release compressed air that would work the signal.

Westinghouse started a business in power transmission, where the problem at the time was direct current. In 1871 electricity was produced by rotating a ring carrying copper-wire coils between the

181 26 28
181 74 50

poles of a magnet made of an iron bar, itself coiled with wire. In this self-exciting, "dynamic" system (from which the name "dynamo" comes), as the coil turns, it sets up current in the wire round the iron bar, making it a magnet. This in turn sets up magnetic lines of force. As each side of the spinning ring cuts the magnet's lines of force going in opposite directions, its coils generate current flowing first in one direction, then in another. This current was made unidirectional by using a device (a commutator) with separate contacts picking up and passing on first one current, then the other. Direct, unidirectional current was the result.

The trouble was that low-voltage, direct current (needed to power lights, for instance) could be sent only about a mile, down thick, expensive copper wire, and a generator was required for every mile the current went. However, without a commutator, the system produced alternating current. This could be sent down thin, cheap wire at very high voltage. If, at the source, a transformer stepped up the voltage (the "pressure" of the current) so it would go long distances, and at the receiving end other transformers stepped the voltage down again for local use, alternating current could free electricity from the constraints of geography. Since electricity could also be made by spinning the magnets instead of the wire coils (which, at high speed, often flew apart), what was needed was a high-speed power source to spin the magnets fast enough to produce high voltage.

In 1888 Westinghouse met that pigeon-loving Croat, Nikola Tesla, who had invented just such a system using alternating current. His idea was brilliant. If the current were sent in separate, sequential bursts — into two or more sets of coils wound on iron — a series of currents out of phase with each other would be set up. As each burst of current happened, around the ring the bursts would create a sequential series of magnetic fields. The magnetic fields would, in effect, rotate — and this could be used to make a magnet spin at virtually unlimited speed. By 1895 Westinghouse had designed a complete generating system around Tesla's idea, winning the contract for the giant new power station at Niagara Falls.

Tesla's motor would also turn out to be just what battleship captains were looking for, as the American navy entered World War I. Thanks to recent advances in metallurgical techniques, new, giant, all-steel, armorplated heavyweight battleships had been introduced by the British, who in 1906 launched the HMS *Dreadnought*. The "dreadnoughts" were also the first "all-gun" turbine-powered warships, and their firepower was extraordinary. New gun-making techniques, using wound steel wire, permitted the construction of longer barrels needed to match the accuracy with which new, metal-jacketed shells could now be fired, because of new, slow-burning explosives. With piston-operating shock absorbers preventing recoil, the guns could also be quickly breechloaded from behind. And because of new, smokeless powder, there was neither flash nor smoke to give away the ship's position or obscure the gunner's aim.

By 1917 the American dreadnoughts USS *Delaware, Utah, South Dakota, Florida, Arkansas,* and *Wyoming*, each displacing 226,000 tons, carried fifteen-inch guns (mounted fore and aft) that could throw a two-thousand-pound shell nearly fourteen miles. Accurate fire control now became an urgent necessity, because at the extreme range these new guns permitted, the slightest pointing error meant a miss. The solution to this problem was, fortunately, already being developed by Elmer Sperry, a New York electrical manufacturer who had created[182] the hi-tech version of a children's toy and had failed to interest Barnum and Bailey's Circus in equipping a high-wire wheelbarrow with one. By 1912 Sperry was in the process of installing several multi-ton versions of his new toy on ships, including the Italian luxury liner *Conte di Savoia*, which in consequence became one of the smoothest-sailing ships afloat.   182 120 *90*

His amazing new device was called a gyroscope. In principle much like a child's top, the gyroscope consists of a spinning wheel mounted on three-way gimbals. The effect of inertia means that as long as the wheel spins, it will always point the way it was originally set. Given the problems that the metal construction of the new, all-steel ships created for any magnetic compass on board, a gyrocompass was a godsend. And thanks to Tesla's electric motor, there was now an electrodynamic means to keep the gyro wheel spinning indefinitely.

Elmer Sperry installed in ships a master gyrocompass that controlled all other compasses on board. He also designed equipment, using the gyrocompass readings, that would track the relative location of a target, would show the ship's own position at any time and would provide the gunners with an arrow showing how they should point their guns. The ability of the gyro to sense the ship's movement on a rolling sea and send compensatory signals so as to keep the guns correctly aligned also meant the ships were now a stable platform, no matter what the sea conditions. Gyro-controlled guns on the USS *South Dakota* shot down every aircraft during an attack in heavy seas.

Sperry's gyrocompass is the second of this chapter's three modern inventions, originally derived from the discovery of the vacuum.

The third of these modern inventions emerged because the seventeenth-century discovery of the vacuum got people excited about air: what it is and, above all, why it is essential to life. Respiration of all kinds came under the scrutiny of English Royal Society amateur scientists. One of them was a Cambridge-educated clergyman living in Teddington, on the River Thames, just outside London. His name was Stephen Hales, and in 1709 he took up the perpetual curacy of Saint Mary's Church. This job-for-life meant that he could devote all his spare time to his two obsessions: how plants breathe, and what makes muscles move. Hales was convinced that the answer to both conundrums lay in the movement of fluids in plant and animal bodies, so he started by looking at why sap rises.

In his Teddington greenhouse, over a period of ten years he stuck glass tubes into plants at various points and watched how far up the tube plant sap would rise at various times, and in various weathers, day and night, year-round. He measured plant-leaf areas and stalk diameters; and then he sealed them in their pots, fed them measured amounts of water and weighed them to see how much water loss occurred through transpiration and evaporation. He grafted onto a plant a tube containing water sitting on top of mercury, measuring the mercury rise as the plant took up the water. What most excited Hales was the force of the sap. Sticking a freshly cut vine stump into one of his glass tubes, he was amazed to see the sap eventually rise to twenty-one feet up the tube. What, he asked himself, was the force driving the sap? And was it the same force that moved blood around the body?

After unspeakable experiments on the various arteries and veins of deer, oxen, mares, dogs, cats, rodents and assorted livestock, Hales worked out that if a glass tube were to be fixed to a man's carotid artery, and the artery were severed, the blood would go 7½ feet up a tube. Hales thought the heart's left ventricle measured fifteen square inches, so a 7½-foot rise indicated that the heart was delivering 51½ pounds of force. In the long run, Hales came up with the first basic book, used by doctors ever since, on blood pressure and how to measure it.[183] Regrettably, he found he was mistaken about how muscles are moved. There just wasn't enough blood pressure in the capillaries going into the muscles to produce the kind of force that muscles exert.

    183 277 *239*

However, not long afterward, at the northern Italian university of Bologna, Luigi Galvani, the Professor of Anatomy, went looking for the same force. Now, in 1780, most people thought if the force wasn't blood pressure it was at least some kind of invisible fluid pumping down from the brain and causing the muscle fibers to become "irritable," so that they contracted. If this were the case, was the force some kind of electricity, the latest wonder-force? Electric shocks seemed to cause everything from flashes before the eyes to nose bleeds, convulsions and jumping in the air. Galvani[184] believed the "fluid-along-the-nerves" theories were wrong, because if a nerve were cut, electricity still flowed through the local area. But in the newly discovered torpedo electric fish, only certain parts of the fish gave a shock. So the force, whatever it was and however it acted, could be confined. And if it could be confined, it could be directed (to muscles, for instance).

    184 99 *69*
    184 216 *179*
    184 234 *196*

Galvani started on frogs and through long, hot, thundery Bologna summer days (it can get really long and hot and thundery in Bologna), he subjected pairs of frog legs, attached to their spine (and cut to reveal the main nerves), to various shocks.[185] He noted that they jerked when his electricity machine sparked and when there was lightning around, but only if his scalpel were touching the frog nerves at the same time. This was inexplicable.

    185 187 *149*

As the days drew soporifically on, the lightning flashes became infrequent enough for things to get really boring, so, almost offhandedly, Galvani began to scrape the copper hook on which his frog legs were hanging against the iron of the stand on which the experiment was mounted. The legs contracted. And they did so every time he scraped. In 1791 a long, rambling and confused Latin monograph from Galvani announced he had found the source of the force. It was in the frog. Animals obviously made electricity. What else could account for the phenomenon?

A contemporary Italian electricity freak, Alessandro Volta[186] (at yet another one of those very busy crossroads on the web, thanks to the fundamental role he plays in history) had the answer. The force was the frog all right, but only because it was playing the role of anything salty and wet (which all organic cells are), so the two different metals Galvani was using (copper and iron) were reacting through this wet, salty frog and producing electricity. Volta proved this by repeating it by another means. He connected a series of cups, filled with a brine solution, via links made alternately of zinc and copper, and got a shock from the final wire link. And the greater the number of cups, the bigger the shock. To make this arrangement more portable for experimenters, Volta stacked discs of zinc and copper, sandwiched between discs of cardboard soaked in brine. Bringing together the ends of wires attached at each end of this pile of discs (Volta called it his "pile") caused a spark to happen. Thanks to Galvani's wet frogs, Volta had produced the world's first battery.

It was not for another fifty years that people actually began to understand what was happening in the salty liquid to produce the electric current: the molecules in the salty solution were adding electrons to one of the metals (the cathode) and taking electrons from the other (the anode). This interaction caused a current to flow in the link (the salty liquid) between the metals. But in doing so, the passage of current gradually dissolved one of the metals. In 1889 a couple of Germans called Mond and Langer found a way to avoid dissolving the metal, making possible an efficient battery. Ironically, Tesla's dynamo killed the idea, and it was not revived again till World War II.

Today, the device has come into its own — thanks to our ability to liquefy gases — in a form that replaces discs and wires with porous electrodes placed in the solution. From liquid-gas sources, one electrode is provided with oxygen, and the other with hydrogen. The hydrogen diffuses through its electrode, which absorbs the hydrogen atoms. The atoms then react with the solution to make water and, in doing so, give up electrons to the electrode, in the form of current. The water and the current are passed to the other half of the setup, where the electrode is busy absorbing oxygen. This electrode reacts with the oxygen (and the water from the other electrode) to create hydroxyl atoms that then migrate through the water to the hydrogen electrode, completing the circuit. The product of this system (called a fuel cell) is electricity and water. This is the third and final modern invention derived from the original discovery of the vacuum.

And, together with the other two inventions (the gyroscope and ceramic tiles), it helps to make possible a fourth invention. First, this needs a constant and autonomous supply of electricity and water provided by a fuel cell. Second, it has to have a gyroscope that will give direction and position unaffected by magnetic or gravitational influences. Third, it needs ceramic protection against temperatures as high as 3,000°F (1,662°C).

All three inventions work together. The fuel cell powers all the onboard instruments; the gyroscopic inertial navigation-and-control system enables the vehicle to orbit the Earth before its pinpoint, one-attempt-only landing; and the crew survives the fiery furnace of re-entry thanks to the 34,000 ceramic thermal tiles covering the belly of the fourth technological marvel: the space shuttle.

In preparation for the rigors of their mission, astronauts are probably the most intensively tested people on earth. Chief among their attributes is the psychological aptitude they have for handling the extreme physical and emotional stresses of space flight. They are, for instance, unlikely to become hysterical under pressure. This is ironic, since the psychiatric expertise of their NASA evaluators originally *began* with hysteria. . . .

11 Sentimental Journey

THIS chapter goes on a journey in search of the origins of that feeling people often get when they're feeling low. They just know they need a break, so they reach for a map and start to plan an exciting holiday that will help take them out of themselves. And the strange thing is, the map is there because they just know they need a break.

It all started with Freud. The reason you know you're feeling low is because of what Freud got up to in Vienna when he changed the way we think about the way we think about the way we think. Freud spent from 1876 to 1882 as an underpaid neurologist dissecting crayfish (which have big neurons that are relatively easy to study) at the Psychological Institute in Vienna. The institute was run by Ernst von Brücke, the leader of the biophysical movement, whose supporters believed that all organisms were governed by straightforward physiochemical forces. The movement had emerged in the early part of the century, in reaction to the Romantic nature-philosophy that had dominated German thought for decades and that proposed vague links between nature and humankind. Biophysicalists, rejecting this view, believed all psychological disorders were simply diseases of the brain.

When Freud got married — and began to worry about the fact that he couldn't support a demanding wife on a neurologist's pay — he decided to switch to medicine and, in particular, the business of worry. In 1882 he joined the staff of the General Hospital of Vienna's Department of Nervous Diseases and started working with Josef Breuer, one of Vienna's most eminent physicians. Breuer had just finished treating a patient (file name: Anna O; real name: Bertha Pappenheim) using hypnosis. Bertha was a highly intelligent twenty-one-year-old who had enjoyed perfect health until suddenly she started suffering from sleepwalking, squint, fading vision, paralysis, muscle contractions and general twitches.

Before meeting Breuer, Bertha had taken all the usual treatments, chief among which was electrotherapy, during which low-voltage current was applied to the patient's limbs. The technique went all the way back to the Greeks, who had placed electric fish on the foreheads of headache sufferers. Breuer thought Bertha's real problem was that she missed her late father. During the two years of hypnotic treatment, he found that symptoms could be relieved by talking to the patient; and, in the end, he managed to cure Bertha in this way.

Freud, however, was more interested in anatomy and set off for Paris (the medical center of the universe at the time) to study the latest treatment for nervous disorders. These methods were less than impressive, since they consisted principally of hosing patients down with cold water and giving them the inevitable electric shocks. Not surprisingly, none of it seemed to work. Disillusioned, Freud returned to Vienna and took up with Breuer again. Eventually, in 1895, they would cowrite *Studies on Hysteria,* the book that founded psychoanalysis. Meanwhile, both of them continued to get patients to relax and talk about themselves. After a while Freud changed his approach. Many of his patients were cases of hysteria (the prime nervous disease of the time) and with them, Freud began using free association. The patient just talked about whatever came to mind, while Freud applied a gentle pressure to his or her head. One of the treatments Freud had come across in Paris was electroconvulsive therapy, still used in some cases today. Nobody knew how it worked, but that didn't matter much back then.

The buzzword in science at the time was "animal magnetism," a phenomenon that burst onto the research scene in the late eighteenth century, when Luigi Galvani discovered what he thought were electric forces being generated by the muscles in a frog's legs.[187] The term "animal magnetism" was not used as it often is 187 185 *145* today, in the sense of generating sexual attraction. It referred specifically to a mysterious magnetic force, flowing from person to person, that could be harnessed as a cure for nervous problems. The last word in nervous research at the time was a converted gunpowder store in Paris, called the Salpêtrière (French for "gunpowder store"). This had been turned into the biggest mental institution in the world, after a spell when it had been used as a reformatory for prostitutes.

By the eighteenth century it housed up to eight thousand mentally ill, deformed, senile or incurable women and girls, as well as quite a few with nothing that Freud couldn't have cured. A doctor-patient ratio of 1:500 didn't help much, and many women resorted to spectacular fits of hysterics just to get a doctor's attention. This was not very difficult, given the medical profession's obsession with the condition. Hysteria had for centuries been regarded as a symptom originating in the womb (in Greek, *hystera*) and, therefore, a disease which could only afflict women. It was also extremely fashionable.

The top man in hysteria (and at whose feet Freud studied) was an unusual character whose nickname was "the Napoleon of Neurosis." In 1862 Jean-Martin Charcot had been appointed head physician at the Salpêtrière and had turned his hand to hysteria. By 1876 his public demonstrations of his theory of "grand" hysteria (four stages, manifesting themselves with exact regularity) were packing audiences into the lecture theaters. There, his most beautiful patient, Blanche Wittman, summoned up genuine hysteria at his Napoleonic command (and was painted by Brouillet doing so).

Charcot identified the root cause of the condition as stress arising from industrial working conditions. These symptoms (shared by both men and women, but causing hysteria only in the latter) were caused by what Charcot called the "jarring" nature of life. This included conditions brought about by traumatic shock (like an accident or violence) combining with a generally weak disposition to disrupt the nervous system. An example of this kind of stress was the trauma Charcot referred to as "railway spine," contracted by riding on bumpy trains.

Charcot treated most things with deep hypnosis, and he showed Freud how to do it. Often his treatment involved the use of magnets, partly because of the influence of the "animal magnetism" school and partly because of a flamboyant type called Franz Anton Mesmer (from whom we get the word *mesmerize*). Today's modern, showman-hypnotist patter ("Now you are feeling sleepy and your eyes are heavy, and when I snap my fingers, you will awake and remember nothing") all started with Mesmer. He was born in Austria and by 1766 had qualified in medicine at the University of Vienna, in 1768 marrying a wealthy widow. He then became a member of the Vienna jet set, befriending the pop stars of the day: Gluck, Haydn and Mozart. The first production of Mozart's opera *Bastien und Bastienne* took place in Mesmer's garden, and later Mozart[188] 188 296 *261* wrote a mesmerism scene in *Così fan tutte.*

Mesmer offered an appealing alternative to the standard, highly uncomfortable contemporary medical practices. He stroked his patients while relaxing them with a soothing, hypnotic voice in a darkened room. During sessions he habitually wore flowing robes and feathered hats. A course of therapy also included immersion in his magnetic bath. Grateful patients often joined his Society for Universal Harmony, although the feeling of gratitude was not always shared by the parents of his patients. One such case was the reason for his eventual hurried departure for Paris. He had been treating a beautiful and talented young aristocrat, Maria von Paradies. Blind since the age of three, she was a brilliant concert pianist, whose father was a court chamberlain and not without influence in Vienna. So when Mesmer cured his daughter's (psychosomatic) blindness, but in the process caused her to lose her musical talent, Mesmer felt it an opportune time to leave town. Whereupon Maria went back to being blind and musical once more.

Mesmerism's "animal magnetism" approach was strongly influenced by the Romantic idea that a mysterious, invisible force flowed through all nature. Even hardheads like Benjamin Franklin thought electricity was an invisible liquid, which explains the origin of some electrical terminology still in use today: *current, electric flow* and *streams of particles.* Mesmerian treatments were supposed to unblock obstructions and free the passage of this fluid through the nervous system.

This concept of fluid had been part of science since William Gilbert had discovered magnetism in 1600. In 1664 Descartes had proposed that streams of "etherial matter" flowed along the axes of magnetic fields through special channels. Even Newton had thought that gravity was one of these "intangible" fluids. So the idea of a universal liquid moving in the body and carrying the vital force was well established when, in 1791, Franz Joseph Gall, another Viennese doctor, published a paper on his theory that the brain consisted of a number (twenty-seven, later amended to thirty-seven) of separate sections. Each of these sections (or "organs") controlled a different part of the intellectual and physical functions by means of tubes of invisible fluids, flowing from each organ to the relevant part of the anatomy. Gall also reckoned that the shape of the skull housing these organs was affected by their size and shape and that the character of a person could therefore be read by the bumps in the skull made by the organs beneath. Thus, for example, a lump behind the left ear indicated that the person was a good lover.

By 1815 Gall had acquired a partner called Spurzheim, and their bump reading (now known as phrenology) was enjoying wild success among the crème de la crème of opinion-formers, including Queen Victoria, Bismarck and President Garfield. Early in the history of phrenology, most of the believers were men of science, but after the 1820s the emphasis shifted to its social potential. Reform-minded middle-class intellectuals and social activists saw phrenology as a tool which could be used to ameliorate the conditions of the poor and destitute, by revealing in them talents whose use might improve their circumstances. Phrenology also provided "scientific" backing for all kinds of wild ideas about women, savages, the proletariat and the education of children. Above all, phrenology might make more likely the emergence of an "improved" and more docile working class. This spawned much self-improvement literature (based on the idea that bump reading could reveal talents that could be improved by study), and courses in the subject became extremely popular in the United States. Also, when photography became generally available, those interested in their bumps could send a picture of their skulls away by mail and have the bumps read long-distance.

Meanwhile, phrenology was to change the world of the criminal. As the material wealth of the newly industrial manufacturing society left more and more possessions and money lying around, more and more robbers helped themselves. The rising crime rate spurred the invention of the police, in 1829, when the first London Metropolitan "bobbies" (so called after their founder, Sir Robert Peel) appeared in distinctive blue coats, Wellingtons, black hats and scarlet waistcoats. However, the only problem with apprehending criminals was finding somewhere to put them. So prison-building programs were introduced.

Previously, prisons had been little different from lunatic asylums, where inmates were shut up and forgotten. But thanks to new social attitudes born of phrenology and the recognition that even criminals might have hidden qualities, things began to improve. Penitentiaries and reformatories were built, as enlightenment began to dawn on the world of incarceration. In 1836 the prison in Cherry Hill, Pennsylvania, was the best of the new design, incorporating cells for individual prisoners. This concept was inspired by a Quaker idea that prisoners kept in solitary confinement to meditate on their sins would more readily reform their characters. Solitary confinement prisons were also especially popular in Germany.

A British libertarian, Jeremy Bentham,[189] designed the "Panoptikon" prison block, with individual cells set around a central control tower from which guards could monitor every inmate's activity. This kind of structure was now easily built thanks to the new availability of high-quality steel with which to construct the walkways, cell bars and tiered bunks. The Panoptikon idea caught on and was widely copied. Reform, rather than retribution, was the aim at new institutions like Sing-Sing in New York.

189 6 *14*
189 14 *22*

It may be because the authoritative text on penology was a best-selling essay written in 1764 by an Italian, Cesare Beccaria, and circulated freely among European intellectuals, that the next step in the story was also to be taken by an Italian. With all the opportunity new prisons offered to study large numbers of criminals, not surprisingly the discipline of criminology emerged. Because of it, phrenology began with heads and brains. Back in Vienna, one of Gall's pupils had noticed that convicts had "peculiar faces," with "marked protuberances . . . large and receding heads, enormous jaws and masticatory muscles always in motion."

By now Darwin was complicating the issue with the theory of evolution and ideas about humans being distantly descended from the apes. Some of us, however, were clearly less distantly related to primates than others. With this in mind, the more an Italian head-measurement fanatic, Cesare Lombroso, measured heads (over a number of years he examined three thousand soldiers and four thousand convicts), the more he became convinced that all criminals looked like apes. In 1870 he had autopsied the famous bandit Villela and found a remarkable depression in his head which was just like that found in lower primates. During a period as director of the lunatic asylum in Pesaro, Italy, Lombroso began to study cretins and criminals, hoping to pinpoint telltale signs of "degeneracy." He even examined the busts of Nero and Messalina for evidence of this trait.

In 1876 he came up with the concept of "atavism," according to which criminals (with "sticking-out ears, hairy heads, straggly beards, giant sinuses, big square jaws, broad cheekbones, sloping foreheads and frequent gestures") were identified as throwbacks to an earlier stage of evolution. Lombroso's book *Delinquents* was a runaway bestseller. It kicked off subclassifications: kleptomaniacs had big ears, deadly females had short arms and criminals were "chronically ill." But it also introduced many modern criminological methods, such as classification of blood, hair and fingerprints.[190] Lombroso's work achieved tremendous popularity, perhaps because he distanced ordinary people from criminals, now defined as subhuman, separate and detached from normal society.

190 271 *229*

True to his beliefs, Lombroso had himself decapitated after death, and his pickled head preserved in a jar at the museum he founded in Turin. He is still there, together with the rest of his collection of pickled brains, variously labeled "German murderer," "Italian bandit," "suicide by firearm," "baby-murderer," etc. But all this aside, Lombroso did have a profound effect on the whole approach to crime and social welfare even as late as the mid–twentieth century, directly influencing the work that led to the famous XYY chromosome debate about whether or not criminality is a genetic trait.

In 1865 into Lombroso's lab came a physician who was to change everybody's understanding of the head and put the bump readers out of business. His name was Camillo Golgi, and Lombroso got him all excited about brain function. (Lombroso also ran a psychiatric clinic.) In 1872 Golgi found himself in financial difficulties and had to leave Lombroso, to take a job as doctor in an asylum in Abbiategrasso, near Milan. There, by a stroke of good fortune, the rules required that he carry out anatomical dissections (he did them in his kitchen-cum-lab, at home). His uncle-by-marriage was one of the first people to use a microscope in pathology work, so Golgi was able to borrow the instrument to look closely at the brain tissue he was slicing up.

Sometime during the following year he made an astonishing discovery. He left a slice of brain to harden for a few days in Müller's fluid (a mix of potassium-ammonium bichromate mixed with osmium chloride), after which he soaked it in silver nitrate solution. This might have been absentmindedness, or more likely he'd heard about the amazing new practice of photography.[191] The Englishman Fox Talbot and others had shown that under certain conditions silver atoms would react to light and produce images. Golgi's silver nitrate did so, too. Upon cutting his sample, he was thrilled to see that the silver nitrate had deposited itself in the tissue — but in a dramatically selective way. Golgi saw the mass of brain tissue had stained a transparent yellow, against which the outlines of brain cells (star-shaped, triangular and rodlike) were clearly visible in glorious black. Golgi had discovered "Golgi cells" (the most prolific cells in the brain). Thanks to this extraordinary experiment, Golgi was able to make some major new statements about the brain. What he said forms the basis for all neurophysiology today: that the neurons are interlaced with one another rather than placed individually on a network and that impulses are transmitted by the overall structure, not just by the individual bits.

The idea that tissue might be stained at all came from the recently established synthetic-dye industry. Aniline dye, discovered by the English chemist William Perkin,[192] was the first of a whole series of artificial dyes to be derived from coal tar. For medical researchers, the key thing about these dyes was the way they clung to bacteria. The man who first made this fact public was Paul Ehrlich, a German researcher working with his fellow countryman Robert Koch, who spent his life in remote parts of Africa finding and identifying the bacilli of anthrax, cholera,[193] tuberculosis, syphilis[194] and other germs. Koch's work was immensely aided by an accident. Ehrlich left a tubercle-bacillus-tissue culture overnight on a warm stove. In the morning he discovered to his horror that some aniline dye had found its way into the culture and, worse, that the stove had been lit. Then he found that the extra heat had stained his bacteria culture bright blue, making the investigation (and, in the long run, the cure) for tuberculosis that much easier to accomplish. Ehrlich had invented bacteriology.

191 44 *34*

192 63 *45*

193 151 *119*
194 176 *138*

One of the things Ehrlich came across, as he made cultures colorful, was that some dyes were also highly successful bug-killers. Methylene blue, for instance, worked as a general analgesic in neuralgia. So when he found that it also stained malaria parasites, he tried administering the dye to a couple of patients. It killed their parasites. From this, Ehrlich developed the concept of chemotherapy, the "silver bullet" approach to treatment, in which a drug is used as a kill-specific weapon that will destroy a particular bacterium and leave the rest of the organism unharmed. Ehrlich's first silver bullet was the cure for syphilis (and the first of the wonder-drugs, salvarsan). This was to get him into trouble with the synod of the Russian Orthodox Church, who reckoned that since syphilis was a punishment from God for an immoral act, the disease should not be taken away with the aid of medicine.

Of all the colors Ehrlich used, the most successful appeared to be methylene blue, because it was a so-called vital stainer. In other words, it could be injected directly into living tissue without causing harm. And unlike most dyes, it diffused through the tissue rather than concentrating around the point of injection. Ehrlich was able to make sue of methylene blue, thanks to the work of another fellow German, Heinrich Caro, one of several chemists investigating what artificial colors could be derived from coal tar.[195]

195 62 45
195 140 108

One reason for the intense German interest in dyeing and chemical work was the fact that the country was not yet united, so a researcher could take out a patent for work in a local principality or dukedom — even if somebody else had a patent for the same thing elsewhere in the country. And due to the fact that the English thought it was ungentlemanly to become involved in trade (heaven forbid — a successful scientist would retire to the country as soon as he had made enough money), by the end of the nineteenth century, the German chemical and pharmaceutical industries had overtaken the British.

One of the new German industries was the huge firm of BASF; by 1868 Caro became its director. But in 1859, as a young married man just returned from England with his English wife, his first work in Germany was at the lab of a man who made his mark on every schoolchild's chemistry lessons. It was in Robert Bunsen's[196]

196 105 72
196 86 59

Heidelberg lab (by this time a world center of chemical research) that Caro made methylene blue. At the other end of the same lab (so to speak), Bunsen and his associate Kirchhoff were busy giving the modern world the means to look into the heart of a star, with the Bunsen burner.

Back in 1846 Bunsen had come up with a way to recycle the gases[197] going up chimneys of cast-iron-making works in England and Germany; then he became interested in all kinds of gases. In 1855 he realized that he could get a hotter flame if he mixed air with coal gas before lighting it.

This also produced a clear, nonluminous flame, free of any minerals burning in the gas. The new clear flame proved to be an excellent tool for analysis because when any material was burned in it, all that appeared in the flame was the burning material.

Kirchhoff told Bunsen that if an intense light were shone through the burning material, the flame would absorb the wavelengths of light given off by the burning material. Then, if the flame were viewed through a glass prism, these missing wavelengths would show up as thin dark lines in the prism spectrum. Kirchhoff and Bunsen had invented spectroscopy.[198] From 1861 onward, it would be possible to identify the composition of the sun and stars — since they, too, were burning.

The spectroscopic phenomenon had originally been discovered by a lens maker named Joseph von Fraunhofer working in a converted monastery near Munich, in 1810. The son of a glazier, apprenticed to learn mirror making and glass cutting, he had originally hoped to make spectacles. Then his workhouse collapsed on him, so he went into the business of engraving visiting cards. That failed. In 1806 he returned to optical work, at a Munich instrument company. Fraunhofer's lifelong obsession was to make optically perfect glass, which nobody had yet managed to produce. Over a number of years he and his French coworker Guinand developed techniques for making glass almost free of striations, by stirring the molten mixture with a special hollow fireclay tube heated to a very high temperature.

In 1814 Fraunhofer was calibrating glass (he was the first to make this a standard technique in lens production) and, as part of the process of inspecting it for the slightest aberrations, used an extremely fine, precise image of a yellow light-line shining through the glass. He had discovered that very fine, yellow lines could be obtained from a flame when it was shone through the glass at a certain angle. So he used this to spot the slightest imperfection in the glass. At one point he was double-checking the purity of a piece

197 110 76
197 240 205

198 86 59
198 105 72

of glass by looking at sunlight, instead of a flame, when he saw more lines. But these were dark. By the time he'd added up all the lines coming from different kinds of light (including the sun and the stars), he found he had no fewer than 574. However, since all Fraunhofer cared about was making pure glass (he was not interested in *why* the lines appeared), when he published his work he kept the commercially valuable glassmaking side of things secret but went into great detail about the mysterious lines. Hence, fifty years later, Kirchhoff was able to develop spectroscopy. But the precision glass manufacture which revealed spectral lines to Fraunhofer also helped him to turn out the first large, high-quality glass lenses for telescopes, enabling astronomers to look into deep space beyond the solar system. With a Fraunhofer lens in his instrument, the great Friedrich Bessel was able to measure for the first time how far away the stars were.

Now, up to the mid–eighteenth century, the only way to see a long way had been with very thin lenses because glass quality was so bad that thinness was the only means to avoid as many imperfections in the glass as possible. However, thin lenses gave long focal length, so some telescopes were hundreds of feet long.[199] The major problem, which thin lenses were designed to avoid, was that of chromatic aberration. Because of refraction, in thicker lenses the blue light coming through the lens would come into focus closer to the lens than did the red light. So at the red focus, stars would have a blue, out-of-focus halo, and vice versa.

Then it was noticed that in the human eye several lenses worked together but did not generate this aberration. So in 1758 the son of a weaver in London, John Dollond, tried to reproduce the same effect by putting different glass lenses together. He put a convex crown glass lens (this had a green tint that cut out the orange-red fringing) together with a concave flint glass, and the two canceled out each other's aberrations. It was this trick that made lenses good enough for Herschel to discover Uranus in 1781.

Telescopic improvements made astronomers and others aware that they were severely limited by the lack of precision with which instruments could be pointed. Things were improved by the marriage of Dollond's daughter, Sarah, to an instrument maker, Jesse Ramsden, who did well out of the deal. As part of the marriage settlement, he got shares in the patent for manufacturing Dollond's

199 131 *98*

achromatic lenses. It was Ramsden who would invent just what Dollond (and all the others) needed most: the means to point instruments with extreme accuracy. In 1766 Ramsden was mass-producing a thousand sextants[200] for the navy, and one of the ways 200 267 224 he was able to do so was with a new technique for making sextant scale markings. This was a fiddly, time-consuming business and difficult to do with accuracy. But an inaccurate scale might send some shipload of navigationally lost souls to a watery grave, so accuracy was a prerequisite. Ramsden's solution was to develop a fine-screw-making lathe which used a diamond cutter linked to, and mirroring, the movements of a sensor tracking along the tiny threads of a fine, handmade screw.

When the finished screw was then mounted at a tangent to a horizontal circular plate (so that the screw threads engaged in tiny teeth cut round the plate's circumference) turning the tangent screw would turn the plate by extremely small amounts. This enabled marks to be made on telescope baseplates or sextant scales, set on the marking plate, with an accuracy of fractions of a second of arc. The invention became known as the dividing engine, and with it Ramsden produced instruments so accurate that they permitted scientific precision never before possible.

Precision balances helped chemists to weigh metals and residues with greater accuracy. Precise gasometers measured gas density, and dilatometers calculated the fine differences in size due to the expansion and contraction of metals at different temperatures. Ramsden's barometers reduced errors of setting by almost ten times. His sextants went on Cook's great voyages of exploration and brought together Dollond's new lenses with his own accuracy of scale. With them, Cook charted twenty-four hundred miles of New Zealand coastline in under six months, a feat considered miraculous at the time. Overall, the precision made possible by Ramsden's new instruments also worked its way into tool manufacture and helped to generate the engineering skillbase needed for the Industrial Revolution to get off to the ground.

But what brings the story full circle is the fact that once the American War of Independence was over, by 1793, enough British troops were released to begin the work which made Ramsden's greatest mark on the modern world. This was a period rife with revolutionary fervor (both French and American), and the British were only too aware of the potential enemy, only twenty-two miles away across the English Channel. It was the fear of French invasion that gave Ramsden his most demanding task, which was to build an enormous dividing engine, called a "circumferentor," with which to make a monster theodolite surveying instrument. In 1783 the national Ordnance Survey of the south coast of England began, using a system known as triangulation. A baseline between two high points of terrain (and running over several miles' length) was measured very accurately and used as the baseline of a triangle, whose side lengths and angles were precisely calculated with Ramsden's theodolite. The sides of the triangle then became the baselines for contiguous triangles, and the process was repeated again and again.

The theodolite had a three-foot horizontal circle made of brass, divided into ten-minutes-of-arc increments, and was readable, thanks to a micrometer, to within one-second-of-arc accuracy. The structure cradled a large telescope and rested on a tripod equipped with spirit levels for all three axes. The instrument achieved a probable error of only five inches in seventy miles.

By 1824 the whole of Britain (except for the eastern part of England and northwest Scotland) had been surveyed in this manner. In 1820 the first agent had been appointed to sell the new maps to the public at the astronomical price of four guineas per county set. Beginning in the mid–nineteenth century, all of Europe and the settled parts of America had maps accurate enough to carry contours and town streets, topographical features and even single buildings. These were the forerunners of the modern map, so useful when you're feeling low and you need to find a getaway spot.

We take maps so much for granted in the modern world, it's hard to remember that there was once a time when the world was largely unknown. And then one day, five centuries ago, Columbus changed our view of the planet when he came home and spoke of a "new world" whose existence had been totally unknown and unexplored. . . .

12 Déjà Vu

T HERE are strange moments in life when you feel as if you're reliving a past experience in every detail. We call that feeling *déjà vu*, and it happens frequently on the web of change, when history repeats itself almost exactly. It does so in this tale, with a sequence of events featuring buried treasure, jungle adventure, brilliant colors, philosophy, swashbuckling buccaneers, great art, music, science and megalomaniacal evil. In other words, this is just your usual humdrum, run-of-the-mill history story.

It begins in the early fifteenth century, when the Spanish arrived in the New World, in what is now Mexico and Peru. Their prime purpose was to find the financial means to restore the fortunes of the virtually bankrupt Holy Roman Empire of Charles V.[201] So in 201 261 *221* 1503, even though it was only eleven years after Columbus's voyage (which few Europeans had heard about, since news traveled slowly back then), Spain had already set up a board of trade to run all American trade through Seville — just in case the New World turned out to be profitable. (As a result, Seville was to become the nerve center of Spanish Atlantic operations.) This "just in case" attitude had colored earlier papal views of the discovery of the New World. In 1493, hardly before the ink was dry on the disembarkation formalities of the returning discoverer, Pope Alexander VI issued the bull *Inter Caetera*, which drew an imaginary line from north to south, one thousand leagues (three thousand miles) west of the Azores, declaring all land and sea beyond it to be under Spanish control.

Further "just in case" decisions were made in 1513 at meetings between Spanish government officials and church authorities in Valladolid and Burgos. It was decided that any American Indians who might be encountered should be required to swear loyalty to the Spanish crown and recognize the Catholic faith, on pain of attack and enslavement. Indians were also officially defined as "inferior" beings, created by God for the purpose of subjugation by the conquering Spaniards. By this definition they were unclean, without civil rights and subject to forced labor (justified as a moral necessity, to cure them of their "natural tendency towards idleness"). And since the church agreed that such enslavement did not contravene human or divine law, Indians were legally unable to withhold their services.

All these decisions were made in light of the Spanish plan not only to subjugate the Americas but to colonize them. One of the earliest chroniclers of conquest, Lopez de Gomara, wrote: "Without settlement there is no good conquest, and if the land is not conquered, the people will not be converted. Therefore, the maxim of the conqueror must be to settle." After a couple of expeditionary visits, by 1521 Cortés had subdued the Aztec leader Montezuma, and in 1532 Pizarro arrived in Peru, with 180 men and 30 horses, to take over the Inca Empire. Pizarro held Atahualpa for a ransom of one and a half million pesos' worth of gold and silver — more than half of Europe's entire annual production of precious metal. Cortés started the process of settlement, with the establishment of new Spanish-style towns in which his men were automatically given town-corporation membership, each with his own land and the right to enslave Indians to help him run it.

Meanwhile, the church attacked paganism, destroying five hundred temples and twenty thousand idols and baptizing more than a million Indians in 1531 alone. But it was the mere presence of the Spaniards that caused the greatest social disruption among the indigenous population. The newcomers had brought with them diseases to which the locals had no immunity. Thanks above all to smallpox, malaria, typhus and yellow fever, by the end of the sixteenth century the Inca population of Peru had fallen from nine million to less than two million. In roughly the same period the estimated decline in Mexico was from thirty million to twelve million.

By late in the century this fall in population was beginning to concern the Spanish authorities, since it was already provoking a serious manpower shortage, and Spain needed all the slave labor it could get for the mines. In 1545 the great silver mountain of Potosí had been discovered in Peru; a year later major deposits were uncovered farther south, in Guanajuato. From 1570 the introduction of a new refining process (using mercury) massively increased production. Silver from Peru and Mexico began to pour into Spain. In the peak years, around 1600, the annual total reached a value of thirty-six million pesos. Between 1550 and 1660 some eighteen thousand tons of silver and two hundred tons of gold[202] were added to European stocks of precious metal. The effect was the opposite of what the Spaniards might have been expecting, because the center of European finance and international trade moved away from Spain to northern Europe. The problem was that by the 1540s the relatively small-scale Spanish industries couldn't keep up with demand from the new colonial settlers. The Spanish Americans needed massive amounts of basic agricultural products and cloth and then, as their stocks of gold and silver grew, so did their desire for all kinds of luxury goods. If these colonists couldn't come home (doing so was forbidden on pain of death), at least they could live well where they were. So their money poured into Europe, causing 400 percent inflation in Spain by the end of the century.

202 263 221

This was nothing but good news for the city of Antwerp, in Flanders[203] (which belonged to the Spanish crown), because by the mid–sixteenth century all the South American capital coming through Seville was arriving in Antwerp to pay for textiles, glasswork, metalwork, art and almost anything else it could be spent on. The general growth in the European economy had already established Antwerp as a major port, sitting as it did on the Scheldt River and ideally positioned between the Baltic and the Mediterranean, as well as between the Atlantic and central Europe. Twenty-five hundred ships and 250,000 tons of freight were arriving in the rapidly growing city every year. In 1500 there were just over 40,000 inhabitants, but by 1568 the number had jumped to 114,000. The city council was obliged to restrict building permits because Antwerp was running out of land. New streets were built, containing smaller houses and gardens that would rent for almost any price to the hundreds of artisans coming to Antwerp in search of Spanish silver.

203 111 80

New World money also brought finished goods to Antwerp

from all over Europe: Italian ceramics, mirrors and glass; French linen and wine; German metalwork and textiles; Dutch and English wool; Portuguese spices; Polish grain; and Flemish tapestries. Extraordinary as it may sound to those who know the sleepy Belgian city today, one sixteenth-century Spanish writer wrote: "Antwerp is the metropolis of the West."

The demand for luxury goods in Antwerp went on rising. Between 1600 and 1670 the number of diamond cutters in the city doubled. A new silk-weaving industry sprang up, as did the production of majolica and sugar refining. Specialist cutters working in alabaster, marble and precious metal replaced traditional wood-carvers. The growing demand from private individuals for works of art stimulated an Antwerp school of painting that included Rubens, Van Dyck and Jordaens. Christophe Plantin[204] ran one of Europe's biggest printing houses.

204   127   95

Not surprisingly, Antwerp's new economic power also turned the city into a major financial center. The growth of capitalism, and the tendency of the time to indulge in nonstop wars, caused a rapid rise in the demand for short-term loans. So for the first time it became possible to make a regular living by lending money on the new, well-regulated Antwerp bourse, the first in the world. Soon the city authorities also guaranteed the security of bills of exchange and deposits. The speculative atmosphere was further enlivened by the introduction of lotteries, maritime insurance and betting.

By now the Netherlands had become the center of Spanish imperial policymaking and military operations; so anytime Spain sneezed, Antwerp caught a cold. Unfortunately, in terms of Spanish royal finances, this began to happen with monotonous regularity. A number of bankruptcies in state-run Spanish entities seriously weakened Antwerp's position. The end came for Antwerp when the Scheldt began to silt up. Then the Protestant Netherland provinces rose in rebellion against Spain, declaring independence as the new state of Holland. The capital city, Amsterdam, was directly on the sea. It soon became the new financial capital of Europe.

By this time the demand for year-round commodity trading had created a pressing need for large, international, properly regulated banks; and in 1609 the first great financial centralizing force on the Continent opened in Amsterdam, at the Exchange Bank. It was the first public bank in northern Europe and was authorized to offer money deposit at interest, as well as to transfer funds between accounts, accept credit and clear bills of exchange. It was also able to exchange money for bullion and foreign coins, so as to mint them into legal tender. Other banks soon followed its example in Middelburg, Delft, Rotterdam and then abroad, in Nuremberg, Hamburg and Stockholm. This new development aided the growth of the new joint-stock companies, initially set up in England to attract investors into seaborne commerce. Suddenly, the business of business was business. The modern financial merry-go-round had begun to turn.

It was round about then that the word *mercantile* began to be heard. (Later, in the eighteenth century, it would acquire the "-ism" used in modern parlance.) What the inflow of Spanish American bullion had done was to put the Western world firmly on the road to capitalism and give it a balance-sheet attitude to life. The bottom line became the only one that mattered. Thomas Mun, in his *Discourse on England's Treasure by Foreign Trade*, dreamed up the idea of a "balance of trade," defined as the relationship between inflow and outflow of a country's bullion reserves. Powerful countries had large stocks of bullion; weak ones had little or none. Bullion paid for investment, industrial growth and armies and was often lent to other countries. Mun's book was translated into most European languages and proved to be a smash hit.

Soon afterward national policy everywhere would be to maximize the import of raw materials by relaxing customs duties on them, while levying export duties on domestic raw materials. Local manufacture using imported raw materials would be protected against foreign finished goods by subsidies and protective tariffs. In 1651 and 1660 England took the next logical step. Two Navigation Acts[205] mandated the use of English ships (whose crews had to be at least three-quarters English) for the transportation of certain specified commodities — for example, sugar, tobacco, indigo, timber and naval stores — on all routes to and from England.

205 3 12

The impetus behind this new way of doing things had come from the Dutch, who had developed a highly versatile and efficient type of ship called the *fluyt*.[206] This innovative and profitable vessel used automated block-and-tackle systems, so it needed only a small crew. *Fluyts* picked up cargoes coming into Dutch ports, particularly from the East, and redistributed them all over Europe. These

206 112 82

imports were at the heart of the bullion issue. The frenzy of European demand for tea, coffee and spices caused a major problem because the Eastern merchants who sold these commodities would only accept payment in precious metal. Later, in the eighteenth century, the British were to find an alternative way of settling accounts, with opium.[207]

207 97 66

Meanwhile, it soon became clear to even the dullest government minister that one way to cut down on the massive bullion drain was to establish one's own colonies. Their only purpose in life would be to provide raw materials for the mother country's domestic industry and thereby also boost her re-export business. For this reason, by mid-century the major northern European powers had set up Atlantic colonies in the Caribbean and North America. The Caribbean was stocked with plantations for growing sugar, and the American colonies (like Virginia) became a source of extremely valuable tobacco. Virginia was a particularly good example of the restrictive nature of colonial laws relating to commerce. The colony was prohibited from trading with non-British merchants or from exporting its goods to non-British ports. This system seemed to work to England's advantage, so by 1650 all foreigners were excluded from any commercial intercourse with any British colony.

The fact that cargoes of dyestuffs were valuable enough to come under the control of the Navigation Acts brings a curious twist in this tale. Most countries had not yet established a permanent navy, preferring to hire ships belonging to private individuals, as needed. These ships, known as "privateers," would carry letters of marque as proof that they were traveling on government business, when necessary. On occasion, however, their crews would indulge in what looked to others like freelance activity. They would attack Spanish galleon treasure ships, kill the crews and make off with the cargoes (or, more often, the ships). But these lawless excesses were — more often than not — covert acts carried out on behalf of government masters, to help the home country's balance of trade. Spanish treasure ships coming back from South America carried a fortune in bullion, pearls, silks, spices and jewels; when the privateers got back home, they would receive a percentage of their haul.

Privateering tactics were so ferocious that they became known as piracy (from the Greek *peiran*, to frighten), and those who committed them were called pirates. Much of their success might have been due to the fact that, as they approached their prey, pirates would indicate their intention of giving no quarter to anybody who might resist, by flying a bright red flag, the Jolly Roger (a name derived from the French *joli rouge*).

One of the most famous of all pirates was Henry Morgan, who in 1655 arrived in the Caribbean on a mission for Oliver Cromwell, lord protector of England. Morgan made his headquarters at Port Royal in Jamaica and then set out, carrying letters of marque, to harry the Spanish enemy. He also slipped in a few raids on the side for himself. In 1667 a peace treaty was signed; but since nobody told Morgan, the following year he took the city of Maracaibo, torturing the inhabitants to get at their treasure. Morgan dealt with the Spanish warships guarding the port by blowing them up with a dummy ship carrying explosives. Booty from this single Maracaibo raid amounted to the astonishing sum of 260,000 gold pieces-of-eight.

For this piece of illegal audacity, Morgan was brought back to London, knighted and returned to Jamaica as lieutenant-governor with orders to arrest and punish all pirates — several of whom he knew and hanged. One of the few of his friends to avoid this fate was William Dampier, who fled to the Yucatán coast of Mexico, where he joined a community of buccaneers. These cutthroats passed the time between raids cutting logwood, a tree particularly prized for the dye-crystals embedded in the wood. On one of Dampier's later pirate missions, he was to capture a ship carrying five hundred treasure chests. From his experience with dyestuffs, Dampier was able to recognize that the ship was also carrying a cargo of cochineal, another dye and every bit as valuable as the jewels next to which it lay.

Cochineal was made from virgin female cochineal beetles, which lived on prickly pear cactus in Mexico and were worth their weight in gold. Prickly pear plantations were specially cultivated for the purpose of rearing the beetles. Production simply consisted of brushing off the plants, immersing the beetles in boiling water and then sun- or oven-drying them. Depending on its method of preparation, cochineal would produce colors ranging from scarlet to crimson to orange. It was this matter of preparation that would turn out to make cochineal one of the most noted colors of the eighteenth century (especially among American revolutionaries).

In 1605 Cornelis Drebbel, a Dutchman who had been apprenticed to an engraver (so he knew about metals) moved to London.

By 1620 his inventions (including a perpetual-motion clock, a magic lantern and an incubator for duck eggs) had earned him quite a reputation. While working on other ideas (a submarine and a microscope), he met four brothers named Kuffler, one of whom married his daughter, and all four of whom became his sales agents. While preparing a solution of cochineal for use in a thermostat-thermometer (one of his other inventions), he accidentally let fall into the cochineal a solution of pewter dissolved in a nitric-hydrochloric acid mixture.

He told his son-in-law of the extraordinary effect this mixture had on the cochineal, and by 1620 the product of the accident was for sale as "Kuffler's color." Drebbel had accidentally discovered how to mordant cochineal. Mordanting was a process by which natural dyes were caused to fix onto cloth fibers. *Mordant* comes from the French word "to bite," because it was thought that the dye "bit" onto the fibers and would not let go. The mordanting process made the color fast; what Drebbel's accident had revealed was that cochineal color could be mordanted by tin, which makes up 94 percent of pewter. Natural dyes change color, depending on the mordant, and in this case the tin caused cochineal to turn a brilliant scarlet.

208 156 *124*
208 303 *277* Ironically, it was Oliver Cromwell[208] (the man who sent Morgan to the Caribbean and gave a boost to the privateering that provided Cromwell with pirated cochineal) who was to make the color known all over Europe and America. In 1645 Cromwell formed his New Model Army and dressed his soldiers in cochineal scarlet, distinguishing the various regiments only by means of different facings on their coats. The New Model Army was the first national military force, enlisted in the direct service of the government and commanded by professional career officers.

With regular pay and good working conditions, Cromwell's soldiers were more amenable to discipline, and in battle they fought well and did not plunder. The idea of dressing them all in the same color might well have been an attempt to induce well-disciplined uniformity of behavior. It might also have been a desire to raise the social standing of soldiery, by giving the men a distinctive look and a sense of regimental loyalty. Either way, it worked. The New Model Army was so well drilled and well equipped that the "redcoats" were the envy of Europe (until they lost the American colonies).

The idea of a drilled and disciplined regular army caught on all over the Continent, but the next major advance in how armies fought had to wait for a flute-playing composer and military genius otherwise known as Frederick the Great of Prussia. In 1740 Frederick inherited the throne, together with the fourth-largest army in Europe. He promptly turned it into the best because he centered all command on himself. He reduced the size of cavalry mounts, so as to create a fast-moving force that would match the mobility of his other major innovation, horse artillery. This new fighting force was equipped with cannons light enough to be pulled on carriages at high speed, and it proved devastatingly effective.

Frederick also introduced fair methods of promotion, good pay and new clothes each year; but in return for these amenities, disciplinary conditions were extreme. Prussian army floggings were legendary throughout Europe, and other forms of punishment included torture, mutilation, branding, solitary confinement and running the gauntlet. The result of these measures was an army of impressive precision and obedience. Every morning soldiers spent a regulation half hour brushing and donning gaiters, and another full hour smearing their hair with wax, fastening it back with a ribbon and liberally sprinkling it with powder. Then followed endless periods of drill, including Frederick's new idea of cadenced marching at standard stride. In 1785, during maneuvers in Silesia, to the stupefaction of observers, a column of twenty-three Prussian infantry regiments wheeled into line, mere seconds after hearing a cannon signal.

These line formations were the product of advances in musket technology. The new model of musket had a conical touchhole that automatically funneled gunpowder into the firing pan, and a new cylindrical ramrod which did not have to be turned round to be inserted. These improvements made possible a rate of fire of five rounds per minute. At this rate, three rows of musketeers could fire and reload in rotation, keeping a constant stream of lead in the air. Frederick's musketeers were the best in Europe, and he never lost a battle.

Frederick's most unusual innovation, however, astonished contemporary commanders. He required that his officers be well educated. Frederick himself was a literate and cultured man, whose admiration for things French (and especially for the work of

209 228 *191* Voltaire[209]) verged on the extreme. After sending the French philosopher letters and poems, in 1740 Frederick finally induced him to come to a meeting. With much haggling on terms, Voltaire agreed to take Frederick's offer of apartments in the royal palaces at Potsdam and Berlin, as well as a pension of twenty thousand francs. The idea was that he should play a star role at Frederick's extravagant suppers, enlivening them with his famous wit. Gradually, however, relations between the two men soured, particularly after Voltaire heard that Frederick had referred to him as "an orange to be squeezed and thrown away." Voltaire then made it publicly known that his real job had been as corrector of the king's prose. As a result of this indiscretion, in 1753 the Frenchman was obliged to leave; and although an eventually apologetic Voltaire continued to correspond frequently with Frederick, the two never met again.

Frederick's interest in things intellectual expressed itself in a more meaningful way when he introduced reforms that would set the basis for much of modern schooling. He standardized primary education with a national curriculum that included prayers, singing and reading of the catechism. He also regulated the hours of teaching, introduced standard textbooks, imposed graduated school fees and instituted fines for truancy. A school inspection service was set up, and teachers were required to write an annual report on every pupil. City schools were brought under the control of the Continent's first ministry of education.

Secondary education was largely freed from the influence of the clergy, and pupils were trained for the *abitur*, the now-obligatory new university entrance examination. Frederick opened the first teachers' training colleges, with rigorous examinations for prospective candidates. He also brought German universities out of the torpor into which they had fallen, with wide-ranging reforms based on Enlightenment principles. And he founded the University of Berlin.

As the Prussian sovereign, Frederick had the right to approve election to many university posts. One of the men he promoted in this way was to have an extraordinary effect on German (and world) history because of his influence over the son of one of the

210 289 *253* royal chamberlains. The new appointee was Immanuel Kant,[210] and he taught physics and math at the University of Königsberg (then in Prussia, today part of Lithuania). Frederick had visited the university when he was crown prince, in 1739, and was singularly un-

impressed. It was, he said, "better suited to the training of bears than to becoming a theater of the sciences."

Kant's major drawback, when it came to influencing his fellow intellectuals, seems to have been that he spoke a virtually unintelligible dialect. But at the age of thirty-one he made his mark with a treatise on the origin of the universe which suggested the existence of starclouds. He also lectured at boring length on physical geography (his favorite topic).[211] His fame in the modern world rests on his philosophical work, the first major publication of which came in 1781, when he was fifty-seven. It was called the *Critique of Pure Reason*, and through it Kant's ideas became widely known throughout the German-speaking world. Unfortunately, because he didn't write in French (the fashionable language of thinkers at the time) it was a while before anybody outside Germany heard of him and his new ideas.

211 260 *219*

In the *Critique*, Kant pulled together the main philosophical theories of the day. The Scottish thinker David Hume had said that all we could know is what we individually perceived through the senses. The German Gottfried Leibniz's rationalism held that it was possible to know the world objectively. Kant argued that while objects might well exist independently of the individual who perceived them, their character was influenced by the observer's viewpoint. And since this viewpoint was constrained by the context in which we found ourselves, these constraints affected the way things were perceived. Kant foreshadowed the world of modern journalistic reporting when he said that the world could be defined by the answers to four questions: *where, when, what* and *how*. Unlike modern journalists, however, the philosopher avoided the vexed question of *why*.

Kant believed that only an empirical approach, based on personal experience, could lead to a true understanding of the world. For this reason, he placed all perception of the world into twelve general categories (or forms of understanding) which would, he said, provide a way to objective truth. Everything that was encountered in life would fit into these categories and subcategories. For example, "knife" was a special category of "tool," in turn a special category of the more general concept "material object," and so on. The twelve categories were organized into four sets: the quantity of objects, their quality, how they related to each other and, finally, their mode of existence.

Kant's ideas were modified by another German thinker named Johann Gottlieb Fichte, in his *Addresses to the German Nation*, published after the humiliating Prussian defeat by Napoleon at the Battle of Jena[212] in 1806. Fichte took subjectivism to extremes, arguing that the ego was a supreme force and that national character was its supreme manifestation. These ideas were to help shape totalitarian nationalism in Germany and would eventually lay the foundations for a political movement that would bring the world close to destruction in the twentieth century. But first, events required the involvement of Alexander von Humboldt, the son of the Prussian royal chamberlain, who had also been deeply influenced by Kant's ideas.

212 313 *284*

One of Humboldt's tutors had translated *Robinson Crusoe* into German, and maybe this story gave Humboldt the sense of adventure that led him on to such an extraordinary life. He began his career in a conventional enough manner, in 1789 enrolling at Göttingen University, where he studied under Abraham Werner, the founder of the theory that all rocks are the product of undersea sedimentation. Werner also taught mining and mineralogy at the Freiberg School of Mines, and by 1792 Humboldt had qualified as an inspector of mines. But then he decided to use everything he'd learned on a great expedition of discovery. It would be one on which, through Humboldt, Kant's ideas on geography (he called it "the basis of history") would lead to the establishment of the subject as a modern academic discipline.

Humboldt's first journey ended before it began. After he waited several years to join a French expedition to the South Pacific, the enterprise was finally canceled because of the Napoleonic Wars. So Humboldt was obliged to shift for himself, and in 1799 he finally set sail from the Spanish port of Corunna to begin a remarkable journey few other explorers have ever matched. After eighteen months traveling around Venezuela, he moved to Cuba for three months, then Colombia, Ecuador and Peru. Early in 1803 he was in Mexico for a year. In 1804 he finally returned to France by way of Philadelphia, where he met and talked to the great Thomas Jefferson,[213] who shared his passion for the environment.

213 80 *53*

213 113 *83*
213 295 *261*

During his epic trip Humboldt shot dangerous rapids, hacked his way through jungles with a machete, came face-to-face with wild animals, lived with primitive tribes and caught various kinds of fever. By the time it was all over, he had accomplished a number of impressive feats. He had collected over sixty thousand specimens, traveled nearly seven thousand miles, found the link between the Orinoco and Negro Rivers, climbed Mount Chimborazo, measured volcanic gas eruptions, established the composition of the atmosphere up to nineteen thousand feet, located the magnetic equator, found the Humboldt Current, discovered guano fertilizer,[214] made over five hundred star fixes and height measurements and, in his own words, "ran around like the demented."

214 24 27

Humboldt also invented environmentalism, as a piece from his prolific writing shows:

> Felling the trees which cover the sides and tops of the mountains provokes in every climate two disasters for future generations: a want of fuel and a scarcity of water. Trees are surrounded by a permanently cool and moist atmosphere due to the evaporation of water vapour from the leaves and their radiation in a cloudless sky. . . . They shelter the soil from the direct action of the sun and thereby lessen the evaporation of the rain water. When forests are destroyed . . . the springs are reduced in volume or dry up entirely. The river beds . . . are transformed into torrents whenever there is a heavy rainfall in the mountains. Turf and moss disappear . . . the rainwater rushing down no longer meets with any obstructions. Instead of slowly raising the level of the rivers . . . it cuts furrows into the ground, carries down the loosened soil and produces those sudden inundations which devastate the country. It follows that the destructions of the forests, the lack of springs and the existence of torrents are three closely connected phenomena.

Humboldt left an indelible mark on history when he wrote up his theories and observations in a giant multivolume work called *Kosmos*. Sales rivaled that of the Bible, and as his reputation spread, tourists flocked to see and hear him from as far away as the United States. Humboldt's was the first theory of vegetation, in which he described plants like communities, driven to cooperative or competitive behavior by the search for food and light. Humboldt conjured up a vision of the tropical forest being like human society, with individual organisms surviving or disappearing in neverending plant wars. He showed how nature and society are both fundamentally constrained and shaped by their environment. Above all, he said, organisms need room to live and multiply. The organisms that do best are those that adapt best to the environment, using that adaptability to expand their territory. When these expansionists meet other organisms, they survive by simply destroying them or by reducing their take of available food. The losers then die, or become slave organisms.

By the late nineteenth century Humboldt's work was inspiring every German geographer. In 1874 his ideas were to be placed in the mainstream of world history, when a travel writer named Friedrich Ratzel took an extensive tour of the United States and saw proof of what Humboldt had said about species and space. In America, the European settlers had moved in and taken the land over from the indigenous peoples. In consequence, just as Humboldt had said, the American Indians were in decline, their economies shattered, their populations dwindling, their lands taken over by white men whose weapons and technology enabled them to adapt more successfully to the environment.

When Ratzel returned to Germany, he quit journalism and took up teaching in Munich, where he began to give lessons in human geography. By 1900 he was teaching in Leipzig — packing the lecture halls with subjects like the scramble for Africa, competitive technology and the need for Germany to establish colonies. In 1901 came Ratzel's greatest work, *Der Lebensraum*, in which he restated a number of Humboldt's fundamental ideas, which were to have a profound effect on his fellow Germans. Ratzel noted that every society and state involves two "ideas," the people and the land; that a superior organism is the one that expands and takes over inferiors, subjugating them to its purpose and taking over their living space by driving them to extinction; that although political boundaries might restrict the territorial growth of a people, political expansion is always a stimulus to racial expansion because it creates more room in which to prosper and multiply; and that history shows how nations either grow through the annexation of other states or die.

Ratzel's book inspired a professor of geography and military science in Munich, who was a retired artillery officer called Karl Haushofer. After Germany's defeat in World War I, Haushofer's lectures on geopolitics and on his concept of the state as an organism gave pseudoscientific rationale to Germany's claims for the return of the colonies she had lost. Haushofer's ideas also inspired his war-time aide-de-camp, Rudulf Hess, who was convinced that the Anglo-Saxon race could dominate the world if only Germany and England could join forces. Haushofer's talk of German expansion was heady stuff, and in 1923 Hess took part in the abortive Nazi putsch. After escaping to Austria, Hess voluntarily returned to Landsberg prison to join the leader of the failed coup, Adolf Hitler. While in Landsberg, Hitler was writing his own political manifesto, *Mein Kampf.*

As Hess recounted Haushofer's version of Ratzel's version of Humboldt's *lebensraum*, Hitler heard ideas that he was, even then, expressing in his own work. The Nazi *lebensraum* strategy was to be executed in several stages. The plan was that after defeating the Allies in 1941, Germany would make sure of raw materials and labor supplies by taking over central Africa, and then large sections of South America. Here, the Nazi conquerors planned almost exactly the same fate for the natives as had the Spanish conquistadors four hundred years earlier.

This particular strategy, of course, never made it past the planning stage. But there is one final irony in this tale of how history repeats itself: the boat that Humboldt had taken across the Atlantic — on the expedition whose ultimate effect would one day bring the Nazis so close to an invasion of South America — was called the *Pizarro.*

Nazi ideology enshrined racism as a basic tenet. "Inferior" races were to be subjugated and used as forced labor. This was an issue that would split the world, but it would not be for the first time....

13 Separate Ways

NO matter how fundamental ideological disagreements may drive people apart, the twisting pathways of history can often unexpectedly bring them back together again. In the case of this tale, the cause of the breakup was slavery — an issue on which, almost from the beginning, opinion was implacably and permanently polarized. Or so they thought.

The enslavement of Africans by Europeans began ordinarily enough when some Islamic Negro captives who had been brought back to Europe by the Portuguese in the fifteenth century offered to buy their freedom with "replacements" from their own tribe. In 1444 the first slave ship arrived in Lagos, Portugal, with 235 of these replacements aboard.

The practice soon became common, and by the time Christopher Columbus arrived in the New World at the end of the century, he made moves to enslave the local population almost immediately. Twenty years later, on the Caribbean islands, African slaves were already at work in the plantations. Sugar (and then tobacco) rapidly became luxury commodities, dependent for cheap production on slave labor. At any time between 1500 and 1800, the way to get rich quickly was to become a slaver. The demand for black labor rose at an astronomical rate, and the English slavers soon became the most efficient and profitable of all. They set up a triangular business, in which slaves from Africa were exchanged for sugar in the Caribbean and then for tobacco in the American colonies.

The transatlantic demand for slaves was almost insatiable. In

1770, one hundred and ninety-two slave ships left Liverpool for the Americas. By 1800 the number had reached 1,283. Profits in the trade were at their peak in Cuba, in 1804, in spite of the fact that slavers had to accept an average 10-percent "commodity loss" (the euphemism for those slaves who died during the voyage) caused by the unspeakably unhygienic conditions on board the ships. With several-hundred-percent profit to be made from each trip, this loss was regarded as an entirely acceptable investment risk. Between 1600 and 1807 a total of 12,420,000 African blacks were forcibly removed from Africa and settled in the New World. The greatest numbers went about equally to the sugar and tobacco plantations of English colonies in America or the Caribbean and to the extensive Portuguese sugar plantations in Brazil.

There were arguments against slavery almost from the beginning. Initially, most opposition to the trade was based on economic reasons. The East India merchants trading in the Far East objected to having to compete with cheap West Indian prices made possible by slave labor. France banned sugar imports in order to protect its own sugar-beet growers. The economist Adam Smith criticized slavery on the (somewhat academic) grounds that it limited the workers' incentive. It was also held that abolishing slavery would reduce the chances of rebellion in the colonies. The British Treasury approved of slavery because sugar taxes gave the government an easy source of revenue. Plantation owners on the Caribbean island of St. Kitts wrote to the English House of Lords to argue that they couldn't cultivate their plantations without blacks, any more than the ancient Egyptians could have made bricks without slaves. John Calhoun, a senator in the American South, held that slavery had advanced the African races to a condition "so civilised and so improved, physically . . . morally . . . and intellectually" that their rivals could not be found among those who had remained in Africa. Calhoun also held the view, common at the time, that slavery was the only means to maintain proper relations between the races.

Gradually, however, the humanitarian view gained ground, and by the late eighteenth century it was spearheaded by Quakers in England and America. Dramatic tactics were employed to publicize the abolitionist argument. In 1738 one American Quaker, Benjamin Lay, arrived at the annual meeting of the Delaware Friends wearing a cloak which concealed the fact that he was in military dress and carrying a sword. He also carried a hollowed-out book that looked like a Bible but contained a pig bladder filled with red juice. At one point during his impassioned speech against slavery, Lay rose and cried, "Oh, all you Negro masters who are contentedly holding your fellow creatures in a state of slavery . . . you might as well throw off the plain coat as I do. . . . It would be as justifiable in the sight of the Almighty . . . if you should thrust a sword through their hearts, as I do through this book!" He then dramatically drew his sword and stabbed the book, spraying nearby Friends with what they horrifiedly presumed to be blood.

The movement slowly gained ground. In a 1772 court case, two lawyers — Granville Sharpe and Will Wilberforce — won freedom for a West Indian slave, James Somersett. Lord Chief Justice Mansfield's ruling in the case declared that under English law no person in England could be a slave: "As soon as any slave sets foot on English territory he becomes free." In 1774 the English Society of Friends voted to expel any member who was engaged in the slave trade. By 1776 the Quakers in Pennsylvania required all members to free their slaves or face expulsion. In 1787 the first English Abolition Society was formed. One of the leaders of the English anti-slavery movement was a Birmingham Quaker called Sampson Lloyd. In those early years of the Industrial Revolution, Lloyd was in the very profitable business of making pins, buttons and nails. In 1765 he was to switch from pins and nails to an even more basic commodity when he founded Lloyd's Bank.

Nail making was a core industrial craft. In the sixteenth century, techniques involved hammering hot metal into rods, heating the rods and drawing them through die holes to make them thin. Then they would be cut, one end hammered flat into a head and the other end sharpened to a point. In 1728 a Frenchman invented grooved rollers that pressed the iron into rods. Sometimes the thinnest rods were wound on drums. But whatever the system, eighteenth-century nail making (and its associated craft, wire drawing) was still a local, small-scale industry; and there were as many techniques as there were producers.

It was a bridge across the Niagara that would change life for the nail and wire makers. In 1831 a German engineer had emigrated from Mühlhausen in Saxony to America, where he founded the city of Saxonburg, Pennsylvania (having refused to settle in the American South because of his views on slavery). He then worked as a farmer, as a surveyor on the Pennsylvania Canal and finally as a railway engineer. His name was John Roebling, and he had a strange obsession with wire ropes. Since nobody in America had ever tried to make that kind of rope, the idea was not easy to promote. After failing to interest the firm of Washburn & Company, in Worcester, Massachusetts (we will return to this firm in our story), in 1848 Roebling moved to Trenton, New Jersey, and set up on his own.

After practicing his technique on a number of small bridges in Pennsylvania and Delaware, Roebling finally got a contract for the new railroad bridge across the Niagara. On-site Roebling spun 3,640 wires into a compact, uniformly tensioned wire cable. Then, using a kite to get the cable to the other side of the river, he went on to finish the first-ever wire suspension bridge, 821 feet in length and strong enough to take the full weight of a train. The bridge opened to rail traffic on March 16, 1855.

Because of his success at Niagara, Roebling's cable-spinning technique soon became standard on all suspension bridges. He put his name in the history books with his next job: the Brooklyn Bridge. When it opened on May 28, 1883, it was considered the eighth wonder of the world and completed the union of the United States from coast to coast. Everything about the bridge was larger than life. It was 5,989 feet long, weighed 5,000 tons and was held up by four 1¾-inch cables, each 3,578 feet 6 inches long. On each side of the river the cables were fastened to wrought-iron plates buried under sixty-thousand-ton granite anchorages, each one of which was eighty-nine feet high. The eighty-five-foot-wide roadway, five feet wider than Broadway, hung from suspension cables.

From 1841 only one other contemporary innovation used enormous amounts of wire the way Roebling did: the telegraph,[215] which would eventually cross the entire country. The trouble was that telegraph engineers needed extended lengths of a particular kind of coarse wire, one that wouldn't need splicing every sixty or seventy feet and that didn't have cuts left by the usual pincer-grip wire-drawing methods used in manufacture. In 1860 an Englishman called George Bedson gave them what they needed when he invented a way to protect and strengthen wire by dipping it in a bath of molten zinc. This technique was known as galvanizing,[216] and

215   30   29
215   147   115

216   184   145

with it Bedson could turn a twenty-five-pound metal billet into a galvanized rod in fifteen seconds. In 1868 a company called Washburn & Company (the same people who had turned down Roebling) brought the new Bedson technique to America, at a continuous-rod mill in Worcester, Massachusetts. Though they didn't know it at the time, the company had set the scene for the transformation of the American West.

217   298   267   Six years earlier the Homestead Act[217] had offered 160 acres of public land free to anybody over twenty-one years of age who would settle and use the land. After five consecutive years' occupation, the property would belong to the settler in perpetuity. However, in many parts of the West this was much more easily said than done, because until 1862, public land had traditionally been taken for free pasturage by cattlemen. And (as the Range Wars were to prove) the cattlemen were not about to give up these rights to farmers without a fight. In the end, the "sodbusters" won, thanks to Bedson's galvanized wire and the efforts of a farmer, a lumber merchant and a livestock breeder. All three men lived in De Kalb, Illinois, fifty-eight miles west of Chicago. And it was in De Kalb, at the 1873 county fair, that they first saw a strange contraption, thought up by a local farmer called Henry Rose. It consisted of a row of nails stuck in a long board, which was set on a fence — with the purpose of keeping the inventor's cow at a safe distance.

Joseph Glidden, the farmer; W. L. Elwood, the livestock breeder; and John Haish, the lumberman, introduced a few modifications to the idea, and in 1874 Glidden patented the finished product. Their version of the invention used two strands of twisted galvanized wire, which they bought from Washburn & Company. The wire was banded at intervals with twisted wire spurs, and they called it barbed wire. It would turn America into the world's greatest grain producer by keeping cattle off the crops.

With barbed wire farmers could fence off their land and, especially, their water (the cause of much of the conflict that ensued with the cattlemen). But the wire brought benefits to the ranchers, too, because it kept expensive imported livestock separate from diseased animals and permitted strictly controlled breeding. At $200 a mile, barbed wire was expensive, but it cut labor costs and reduced cattle loss. And it did away with the need for cowboys. By the turn of the century, because of barbed wire, on the wide-open spaces of Nebraska and Illinois — where once the cattle had roamed — there was nothing to see but acres of waving corn.

This change in scenery was nothing but good news for a couple of New York canners, Merrell and Soule. Merrell (who had the ideas) met Soule (who had the money) in the Syracuse grocery store where Merrell was working at the time. Together they automated the process of putting corn in cans for the fast-growing market in the industrial cities along the East Coast.

By the 1880s the Merrell and Soule "automatic line" machine was cooking corn in a Rube Goldberg device called the Merrell Gun Cooker. This machine consisted of a steam-heated locomotive-boiler cooking tube with a large propeller at one end of it to blow the corn through the tube. At the other end, the corn dropped into cans that were then automatically packed and sealed. By 1890 the Merrell and Soule Chittenango factory was the largest canning operation in the northern states, turning out 150,000 cans of corn a day.

Merrell and Soule's success encouraged them to branch out into canned dried milk powder and canned mincemeat. Then disaster struck the industry in the form of "corn black." After corn had been in the can for a time, it developed black spots. In 1909 research showed that the spots had been caused by chemical reactions produced by the tin-soldering process. Manufacturers tried everything they could think of to solve the problem, including steel cans, thicker tinplate and cans lined with everything from parchment to lacquer to linseed oil. Finally, they enameled the interior of the cans, which did the trick. At first the enameling was done with the old molten-bath technique, invented by Bedson for galvanizing wire with zinc. A more advanced technique, tried in the 1930s, electroplated the cans with cadmium. This material also turned out to be highly effective at rustproofing everything from car parts to refrigerator trays. Cadmium was doubly attractive because it was a free by-product of zinc smelting. Then it was discovered that the metal was highly toxic. And so its use was immediately discontinued.

But this was not before cadmium was found to have one other property, which would turn out to be vital to national security and would help develop the most controversial technology in the modern world. Ironically, the development would also reunite the two, separate historical sequences triggered at the beginning of this chapter by the controversial issue of slavery.

Back in the eighteenth century the abolitionist movement had enjoyed the powerful support of a wire-and-nail manufacturer. At the same time, the slavers were to find their greatest allies among those who manufactured a different, though equally essential, commodity. It was sugar. In terms of its trading value, sugar was so valuable that after one Anglo-French war, serious consideration was given by the English to returning Canada to the French instead of the Caribbean island of Martinique, on the grounds that Martinique had sugar plantations and Canada didn't. By 1800 the sweet stuff represented half of all imported European grocery items. Up to 1830 sugar was made into candy, icing sugar and iced buns. Then came "lollipops," sweet lozenges made with sugar or treacle (molasses). Sugar was also used for coffee and drinking chocolate (the chocolate bar did not appear till 1847) as well as tea. Sugar for daily use was generally sold in the form of sugarloaves.

The production of sugar triggered an almost instant market for a new drink made by distilling molasses. Before the seventeenth century there is virtually no mention of it, but by the early eighteenth century the rising price of the grains needed to make hard liquor had persuaded the market to look elsewhere for a cheaper substitute. It was rum, and it revolutionized every seafarer's life. The Royal Navy even made it their official tipple, issuing every mariner a daily ration. In 1698 only 207 gallons of rum were imported into England, but by 1775 average annual imports exceeded two million gallons.

Between 1660 and 1775 consumption of various forms of sugar in England had multiplied twentyfold. From being a rarity in 1650, sugar had become a luxury by 1750, and by 1850 it was a necessity. Sales grew particularly rapidly once the Industrial Revolution had put spare cash into laborers' pockets, and the workers spent it on sweet tea. Sugar boosted their low calorie intake and added variety to their otherwise bland and monotonous diet.

In 1807, when Britain and the United States finally banned the slave trade, the sugar industry took a massive blow. Imports from the British West Indies fell by one-half. With slavery outlawed, it became urgently necessary to find laborsaving devices that would keep sugar production high and costs low. Fortunately, back in 1813 an Englishman called Howard had invented a vacuum pan, in which liquids would boil well below the normal temperature. This meant that the process required less fuel, and with lower temperatures there was less danger of scorching the sugar as it crystalized out.

Then a black French Louisianan named Norbert Rillieux found the answer to the sugar producers' prayers. As a young man Rillieux had been sent to Paris to complete his education and had qualified there as a mechanical engineer. For a while he had lectured at the École Centrale, before returning to New Orleans in 1834. Nine years later he invented his Multiple Effect Evaporator.[218] Rillieux's evaporator consisted of a series of vacuum pans, each one of which was heated by the steam produced by the previous one. This system offered significant fuel savings; within six years the evaporator had been installed in thirteen factories, producing forty-five hundred tons of product a year and establishing Louisiana as America's premier sugar producer.

218 85 58

The news soon spread. By 1850 the machine was working in Demerera, and the Europeans were using it in their beet-sugar factories. Since there was no international copyright law, in 1850 a French engineer called Cail got hold of Rillieux's blueprints and took out his own multiple effect evaporator patent. Meanwhile, back in Louisiana, Rillieux was ready to leave, tired of the slights to which his African ancestry exposed him. In 1861 he returned to Paris to study Egyptology with one of the Champollion brothers, who had translated the hieroglyphics on the Rosetta stone. Rillieux also went on making improvements to his evaporator and in 1880, when he was seventy-four, produced an even more energy-efficient model.

The multiple effect evaporator worked because steam contains extremely high levels of heat stored in the vapor. This had been discovered in the mid–eighteenth century by Joseph Black, professor of chemistry at Edinburgh University, during his search for ways to reduce the local Scotch whiskey distillers' rising fuel costs. The recent union of England with Scotland had opened the English and American markets to Scotch exporters and, given the severe shortage of wood at the time, the whiskey makers were keen to get more bang for their buck. Black measured how much energy it took to melt a pan of frozen water, and then how much more (a great deal) it took to boil the water away. From these experiments, Black was able to show the distillers the minimum amount of wood they needed to burn in order to evaporate a given quantity of whiskey mash, and then how much cold water it would take to condense the whiskey out of the steam. This research also suggested to his friend James Watt, the technician and instrument maker at the University of Glasgow, a way to make steam pumps work more efficiently.

219 270 227   In 1765 Watt used Black's[219] knowledge about latent heat to
219 304 278   make a minor improvement to the steam pump — and changed the world. The problem with the pump was that when hot steam was injected into the cylinder to push the cylinder piston up, the steam heated the cylinder. Then a jet of cold water entered the cylinder and condensed the steam. This condensation caused a partial vacuum inside the cylinder, and external atmospheric pressure pushed the piston down again. The trouble was that on each cycle the steam was so hot that the cold water would only partially cool the cylinder. So on the next cycle the steam would condense less than
220 170 134   it had done before. This went on until the vacuum[220] no longer formed, and the engine stopped working.

Watt solved the problem by connecting the cylinder via a pipe and a valve to a separate container, kept cold under ice water. When steam entered the main cylinder and pushed up the piston, the valve to this separate container would be opened and the steam would rush in to fill it, too. Here, thanks to the surrounding ice-cold water, the steam would rapidly condense and a vacuum would form. Through the now-open valve and pipe, the vacuum would spread to the main cylinder. But because the ice-cold condenser was separate, the main cylinder would remain hot, avoiding the earlier problem of premature condensation.

This use of latent heat theory gave Watt the kind of pumping power that would drive machines all day. So he went looking for a backer, and in 1768 he found one in the person of William Boulton,
221 17 24   who had a water-powered factory in Soho, Birmingham,[221] where
221 136 105   he employed six hundred artisans making shoe buckles, buttons,
221 305 278   sword handles, watch chains and assorted trinkets. Boulton saw the potential of Watt's machine and lent him the money he needed, encouraging him to lease the steam engines to mines, collieries and (in a version which drove bellows) iron foundries. Within a few years Watt was famous and rich.

There was only one thing wrong with making a fortune back then: you had to take your income in anything but cash. Thanks to the fact that there was no properly enforced regulation, half the

coins in England were counterfeit.[222] Consequently, a lot of private
companies and provincial authorities issued their own coinage. So
in 1786 Boulton turned his button-making expertise to making
money with a steam-powered coin stamper. His first contract was
to make a hundred tons of copper coin for the East India Company,
a happy coincidence since Boulton owned shares in the Cornish
Coppermine Company. By 1788 he had six coin-stamping machines
and was stamping coins for various South American colonies and
the Sierra Leone Company. Impressed by Boulton's work (and his
lobbying) the British Privy Council asked him to submit designs
for new national coins: the penny, halfpenny and farthing (quarter-
penny). At the same time, he also picked up orders from France,
Bermuda and India. By 1792 eight of his machines were striking
medals to commemorate everything from the Russian emperor's
coronation to the execution of the French queen, the Battle of Tra-
falgar, the Hudson's Bay Company and (ironically) the abolition of
the slave trade.

Boulton's one-man machines could strike any number of pieces,
at rates between 50 and 120 a minute, depending on the diameter
of the coin and the degree of relief. Each piece was struck in a steel
collar, so they were all perfectly round and of equal diameter. And
since the automatic machine impacts were more uniform than when
coining was done manually, the dies were less liable to break. In
1797 the designs Boulton had submitted earlier won him the contract for the official English twopence, penny, halfpenny and farthing coins, as well as a commission to build and supply the new
mint, on Tower Hill in London. Typically, Boulton designed everything from the building layout to the coining machines.

In the early nineteenth century the designs on the coins issued
by the new mint, together with Boulton's high-quality stamping
techniques, brought about a kind of nationalist artistic revival in
numismatics. This was when the "Marianne" figure first appeared
on French coins, and "Britannia" on English money. Coin motifs
were heavily influenced by neoclassicism, possibly because one of
Boulton's friends and clients was Josiah Wedgwood,[223] many of
whose pottery designs were for the architect Robert Adam.[224]
Adam had recently returned from Italy (where Pompeii had just
been discovered) and had revived the British interest in ancient
Greek and Roman styles.

222 9 16

223 293 260
223 307 278
224 294 260

In 1824 Benedetto Pistrucci, an Italian engraver employed to design even more new coinage (he introduced the image of Saint George and the dragon on the new gold sovereign), brought another novelty into minting. It was a device called a pantograph, or reduction machine. The pantograph used a series of pivoting arms, each smaller than the previous one, to link a stylus at one end with a cutting instrument at the other. This allowed the engraver to create the design on a large-scale model made of some easily worked material. By mid-century this model was usually made from plaster and, when the design on it was complete, was plated with a nickel surface. The model would then serve as the working copy on which the engraver could trace out the shape of the design with his pantograph stylus. The reduction mechanism would then drive a cutter to reproduce a much smaller version of the same design in the steel of the coin die. Pistrucci's pantograph was so successful that the mint bought a duplicate, and the chief engraver, Will Wyon (who got the job because Pistrucci's foreign nationality disqualified him from the post) used the reduction technique to cut the master dies for Queen Victoria's first coinage.

The deposition of the nickel surface on the working model was made by the new electroplating technique, originally discovered by 225 186 *146* Alessandro Volta's colleague Brugnatelli. Volta[225] had used chemicals to make electricity in his pile (the first chemical battery), so Brugnatelli reckoned things might work the other way round: electricity might cause chemical reactions. He showed that if, for example, you put any material (connected to a battery) in a bath of copper sulfate containing a piece of copper (also connected to the battery), an electric charge would cause copper atoms to be released from the solution and deposit themselves on the material, electroplating it. At the same time as it was giving up its atoms, the copper would gradually dissolve into the solution.

In 1833, after investigating this process, the English scientist Michael Faraday discovered that each depositing mineral, such as copper, needed a different electric charge to make it release its atoms. This discovery indicated that there had to be some relationship between the charge and the mineral which depended on the mineral mass. A different mass would relate to a different charge. From this hypothesis, Faraday worked out his two laws of electrolysis: the amount of chemical change from a current is proportional to the quantity of electricity used, and the amount of chemical change produced by the same amount of electricity in different substances is proportional to their weights.

By the late nineteenth century Faraday's laws enabled the scientific community to look closer at charge and mass, as they related to the recently discovered X-rays[226] and other mysterious electrical phenomena. At Cambridge, in 1910, a researcher named J. J. Thomson was looking at what would happen if electrons were shot through low-pressure gases, and it was while shooting these particles through electric and magnetic fields that he discovered the fields could be made to deflect the particle trajectory. So he tried this trick with neon and saw the stream of particles split into two, as if the neon had two different masses — one lighter than the other and more deflected by the effect of the fields. Atoms like these (of the same element but with different masses) became known as isotopes. By 1919 Thomson's assistant Fred Aston was able to achieve the separation of isotopes of materials whose weight differed by only 1/100,000,000. The new technique was called mass spectrometry.

With a modern mass spectrometer you can identify the constituents of any vaporized material, by firing particles of it through magnetic and electric fields. Where the particles fall tells what they are with an almost incredible degree of precision. This means you can identify minute traces left behind at a scene of crime. Or pinpoint the presence of a steroid in the blood of an athlete. Or monitor the chemical tags attached to a drug while it's in the patient's bloodstream. Or identify infinitesimally small residual amounts of explosives found on suspect terrorists.

With the mass spectrometer you can also take a material from which you require a particular one of its isotopes, vaporize the material, shoot it through electromagnetic fields and collect the isotope you want from its specific landing zone. In the early 1940s this isotopic separation technique would change the world. Two days before World War II broke out, an obscure scientific paper had shown that nuclear fission was more likely to occur in uranium 235 than in uranium 238. The trouble was that uranium 238 was 140 times more abundant in nature than uranium 235. In other words, it might prove impossible to find enough natural uranium to fuel a nuclear chain reaction. There was only one alternative: to separate

226 39 *31*

226 116 *88*

out the rare U235 from the more abundant U238. In a desperate race against the Nazi scientists who were trying to do the same thing, the top-secret Manhattan Project at Oak Ridge, Tennessee, finally hit the jackpot. One of the techniques they used was mass spectrometry. When uranium isotopic separation succeeded, in 1943, detonation of an atom bomb[227] was only a matter of time.

227 49 40

So the two trails in this chapter finally come together. First, the series of events triggered initially by the pro-slavery lobby led through sugar evaporators, steam engines, coinage and electroplating to atomic weapons. Second, the sequence initiated by the anti-slavery lobby led to the Brooklyn Bridge, galvanizing, barbed wire, canned corn and tinplating with cadmium, which ultimately turned out to act as a neutron-absorber in the rods controlling the reaction of the first atomic pile.

Atomic power has for some time been the showcase of those who believe that swords can be beaten into plowshares. Sometimes, however, history will do the opposite....

14 Routes

MAYBE one of the most fascinating things about the grand web of change is that it isn't so grand. Everybody's on the web, from world-altering geniuses to nobodies. Except, of course, on the web *nobody's* a nobody. On the web each one of us makes a contribution to the process of change. Sometimes even the most humdrum event, in an otherwise ordinary existence, ends in consequences that shake the planet.

Take, for instance, the case of Jethro Tull, a perfectly unexceptional, middle-class Englishman who was called to the bar, three hundred years ago, at Gray's Inn, London, and prepared for a life of law. Nobody knows what happened next, but he began to suffer from ill health; so in 1709 he left the courts, bought a farm near the quiet English village of Hungerford and settled down to enjoy the less-stressful life of a gentleman farmer. Before long, his ill health caused him to seek recovery in a better climate, so in 1711 he left for Italy and France. By 1713 he was staying among the rolling green hills of Languedoc in southwestern France, near Frontignan.

It was here that events took a turn for the unexpected. Tull noticed the method local winegrowers were using to avoid the unpleasant taste they thought their wine would have if manure was used on the vines. Instead, they planted the vines in straight lines and deep hoed between them with a plow, at regular intervals through the growing season. This method had the effect of destroying the weeds and keeping the soil turned over. So the vines grew well, without the use of manure.

In time, Tull's health improved enough to go home, and when he arrived back in England he tried the same trick on his own farm. Initially, he limited the practice of hoeing to the land set aside for turnips and potatoes, but this proved so successful that he tried it for wheat. Tull discovered to his astonishment that he could grow the crop for thirteen years on the same land without the need for manure. What was more, the yield improved dramatically. Hoeing increased output (which made money) and reduced expenditure on manure (which saved money). In 1733 Tull published all the details of his experiments in one of those printed cures for insomnia, *The Horse-Hoeing Husbandry*. After initial resistance from the traditionally conservative English landowners, somebody translated the work into French. That did the trick. The English gentry thought all things fashionable began and ended with the French, so suddenly horse-hoeing was chic.

The new technique had one major thing going for it: it came along right in the middle of the great English Agricultural Revolution. This had kicked off nearly a century earlier, with ideas like crop rotation and the introduction from Holland of new plants (like clover) that would nitrogenate the soil. The practice of enclosure was also spreading fast. Fencing common land protected animals from predators and disease, kept livestock healthier and made possible the selective breeding of animals that would produce more meat. More food meant cheaper food, so people married younger and had more children. The surge in population generated a rapid rise in the demand for manufactured goods, which in turn triggered the Industrial Revolution.

It was ironic that in Tull's case the French should have had such a beneficial effect on English agriculture, since their own was in terminal decline. The appalling state of French roads made it impossible to develop a national market, so economic activity was fragmented and commerce was mostly local. The restrictive, feudal nature of French property rights meant there was little incentive for new industrial money to buy its way into land in the same way that had boosted agriculture in England. What few French freeholds there were tended to be small parcels of land, belonging to conservative peasant farmers living at subsistence level.

Even if a national market were to be established, trade was hamstrung by a dozen different regional systems of weights, measures and tariffs. The situation was a catch-22: because of a limited amount of industrialization, growing city populations couldn't feed themselves — and so the authorities were obliged to fix the price of bread. However, to ensure continuity of supply in this unprofitable market regime, the law also forced peasants to sell their produce at market within three days — at any price. So there was little incentive to increase production.

Into this mess came François Quesnay, a country doctor turned surgeon, with a boundless admiration for English land management, who might have risen without a trace had he not been in the right place at the right time and caught the eye of Louis XV's mistress, Jeanne-Antoinette Poisson (otherwise known as Madame de Pompadour). Quesnay's total lack of courtly manners and his uncompromisingly direct way of speaking must have had a certain curiosity value in an era of bow-and-scrape, because in 1749 Quesnay became Pompadour's personal physician and was installed on the mezzanine floor (just below her apartments) of the palace at Versailles. Since his new royal quarters were on the way up the stairs leading to Pompadour's rooms, Quesnay got to meet all the visiting notables and intellectuals of the day, including Voltaire[228] 228 209 *170* (who was, incidentally, a fan of Tull's agricultural techniques). Quesnay's popularity with the royal family was rumored to have stemmed from the fact that he had cured Pompadour's frigidity by taking her off a diet of vanilla, truffles and celery and getting her to exercise.

Quesnay's staircase meetings soon included intellectual luminaries who turned up for Pompadour's regular salons. Among frequent visitors were liberal thinkers like Condillac, Buffon, Helvétius and, above all, Diderot, who was the editor of an amazing new encyclopedia that aimed to present all knowledge in modern, rational form. In 1756 Quesnay wrote two articles for the encyclopedia, one called "Farmers" (1756) and the other, "Rural Philosophy" (1763). In the articles he expounded his ideas on how to cure France of her economic malaise. Quesnay's theory became known as "physiocracy" because it linked the human condition directly to the physical conditions of life. Get agriculture and food production right, he said, and everything else would follow.

Quesnay based his ideas on the belief that the extraordinary advances happening in English agriculture were due to the existence of a national market, where individual producers were free to buy and sell without restriction. The disastrous situation in France proved Quesnay's point: that nature knew its own requirements and that any interference with the natural processes would only destroy the natural order. The results of such interference could be seen only too clearly from Paris to Marseilles. Quesnay's plans for deregulation appealed to land-owning aristocrats, who wanted to be able to sell their products or withhold them from the market, as they chose. However, the fundamental aim of the Physiocrats was to reduce the price of bread, which they considered to be the key factor controlling any country's political stability. "An honest loaf" would save France, but it would come only from the competition that free trade[229] and a laissez-faire approach would encourage. Cheaper bread would generate economic recovery, as it had already done in England. The Physiocrats argued that the condition of the mob in France was so desperate that, if the government did not adopt these new policies, things could turn quite savage.

229 4 13

"Savage" was a new term introduced by philosopher and permanent Swiss exile Jean-Jacques Rousseau, whose influence on European thinking was considered by Napoleon and other movers and shakers to have been primarily responsible for the French Revolution. Rousseau had originally been obliged to leave Geneva for France because of his left-wing politics, spending much of his life flitting between France, England and Switzerland until his death in France in 1778.

In 1754 Rousseau wrote an essay in which he used the term "noble savage" for the first time. In the French world of kings by divine right, aristocrats with absolute power over their serfs, summary justice and bureaucratic corruption on a vast scale, Rousseau looked back nostalgically on the simpler life of primitive man. Probably with the American Indian in mind, he referred to a savage living at one with nature amid the unspoiled beauties of forest and mountain. He described such an existence: "I saw him satisfying his hunger under an oak, quenching his thirst at the first stream, finding his bed at the foot of the same tree that furnished his meal; and therewith his needs were satisfied."

But from this natural, unfettered existence — self-sufficient and with no need of others — humans had then moved into social groups because of their rising numbers and the need to share resources. For organizational reasons, people had then been persuaded to place these common resources in the hands of a sovereign, who would enforce laws that would keep the people safe. Thus had appeared the concept of rights of possession, as well as laws to perpetuate such rights and to deny those without possessions the opportunity to express their views. For Rousseau this process of socialization had caused humankind to degenerate.

In *The Social Contract*, the work that did the most to inflame the republican passion of the French, Rousseau developed the theme. The original, natural freedoms should be every person's right and should be restored. Laws should only be the expression of the wishes of ordinary people, and it should be the people alone who sanctioned any government's existence by their vote. Those who drafted the laws should have no legislative power. Sovereignty was nothing more than the body politic in action and the "exercise of the general will." Above all, the only possession needed to qualify any individual to vote in such a democratic society was the possession of "feeling," the innate touchstone of truth and goodness that all individuals possessed.

These thoughts did not endear Rousseau's work to the royalist government in Paris. On the eve of the French Revolution, his message was being read aloud in the streets of Paris, smuggled copies of his books were everywhere and clandestine meetings discussed and helped to spread his ideas further. In 1789 he was hailed as the intellectual founder of the new republic. But it had taken more than Rousseau's words to bring about the collapse of the French monarchy. The last straw was the disastrous effect on the French economy of another revolution happening on the other side of the Atlantic.

The American War of Independence offered France a great opportunity to do what she had wanted to do for years: damage Britain. The sugar-rich French West Indies were threatened by the British navy; French Canada had been lost to Britain; and with French industry suffering from the loss of Southern cotton thanks to the war, it was in French interests to help the American rebels in any way possible. So an already-famous playwright, Pierre Augustin Caron de Beaumarchais (whose works included *The Barber of Seville* and *The Marriage of Figaro*) was co-opted to arrange matters. Beaumarchais set up a fake company called Hortalez & Company to launder the money going to the American fighting fund. In case

French ships were stopped by the British, he also doctored cargo manifests for French munitions being sent to America. He even hired agents provocateurs to whip up anti-British sentiment among the colonists.

French public opinion (among the literate few) was generally supportive of the venture. Penniless young aristocrats welcomed the opportunity for advancement and money that a transatlantic military operation would offer. The king backed the project because the prime minister, Necker, had cooked the national books to make it look as though France could afford the adventure.

But whatever the cost, France achieved her goal of detaching America from Britain. Thanks to the arrival of the French admiral de Grasse (with thirty ships and over three thousand troops) in Chesapeake Bay in September 1781 to back up the French siege artillery arriving at the same time from the French garrison in Newport, the British in Yorktown[230] were outgunned and outmanned. Their surrender ended the war. But France would pay dearly for having supplied 90 percent of the rebels' guns and filling the American war chest. Between 1774 and 1789, as a direct consequence of the expenditure on America, interest on the French royal debt rose from about $150 million to about $500 million. Added to the problems caused by the king's unbelievable extravagance, this debt was enough to push the state into bankruptcy and financial chaos. The French Revolution was the inevitable outcome.

But the Revolution did more than remove a corrupt political system. Together with what had happened in America, the new democracies gave birth to a different view of the world. Until then, thinkers and artists looked back to classical models for their inspiration. Science had shown, through the work of people like Newton and Leibniz,[231] that the universe worked according to mechanistic laws. The new thinking of the Revolution saw this as further evidence that science and society had become too machinelike, with individuals reduced to cogs in the machine, and knowledge too fragmented and specialized to be socially valuable. The time had come for a new ethos that would be at the same time much more individualistic and concerned with the union of humankind with nature that Rousseau and others had talked about.

The first expression of these new ideas began to appear in Germany at the end of the eighteenth century. The movement was led

230 13 20
231 82 55
231 253 214
231 309 279

by a twenty-three-year-old professor at the University of Jena. Friedrich von Schelling developed what became known as nature-philosophy. He attacked the old Cartesian, mechanistic view of the universe and tried to find a way of bringing all the sciences together in one unified theory. He probably got his initial ideas from K. F. Kielmeyer, a Jena biologist who was lecturing on the sensitivity of living organisms.

At this time it was beginning to look as if experiments with electricity and magnetism would reveal a single, indivisible force that worked together with gravity. The force obviously expressed itself in the life processes and in the will of the individual. So any science that separated the human spirit from nature and its forces was mistaken. Von Schelling's ideas drove the early Romantics[232] to investigate the ways in which individualism expressed itself. What mattered was not a sterile, mechanistic theory of existence but how the senses perceived reality. In music, literature and art, individuals were to be portrayed alone with their feelings, free to seek communion with nature. True knowledge came from the experience of the heart, not the ratiocination of the head.

232 256 *215*

Naturally, all this imprecision got a very mixed reception from the scientific community. It was said among German medical researchers that nature-philosophy was "the Black Death of our century." But ironically, it was in medicine that Romanticism was to have the greatest effect. It was in reaction against the "mystery force" nature-philosophy school of thought that set Johannes Müller, professor of physiology (which he invented) at Bonn, to find out — physiologically — what actually happened during the process of "feeling." In 1840 he came up with the Law of Specific Nerve Energies. Whenever each organ was stimulated, it gave rise to its own specific sensation and no other. The ear reacted only to sound, the eye to light and so on. And it was the organ that determined the reaction to the stimulus, not the stimulus itself. To the organ it made no difference whether the external event affecting it was light or sound or electricity. It could even be that organs were stimulated by internal events, such as the activity of the imagination. This might explain why people saw ghosts. Müller also experimented on animals and theorized that there might be separate motor and sensory nerves present in the autonomic nervous system.

Müller's 1840 *Handbook of Physiology* made a great impression and got everybody interested in investigating the nervous system and how it worked. It may be because Müller had once said, "The will sets in activity the nervous fibers, like the keys of a piano," that

233 87 60   one of his pupils, Hermann von Helmholtz[233] (a surgeon who played the piano all day but didn't like the "new" Romantic music), looked closer at this concept. Beginning in 1856, he wrote a series of papers on sound: how tones are heard and how the mechanism of the inner ear works. Helmholtz was basically looking at an aspect of "feeling," expressed in the awareness of pitch, timbre and loudness, through which a hearer is able to identify one instrument or voice from another.

Helmholtz came to the conclusion that the inner ear contains vibrators, each tuned to different frequencies and exciting the appropriate nerve, so as to transfer into the brain the impulse relevant to that frequency. One day when he was working with a singer, he saw that if a sung note were held long enough, the string on the piano corresponding to that note would vibrate — as did certain others, at intervals above and below the note.

Helmholtz followed up this discovery with a series of sound-related experiments, using an electromagnet to make a tuning fork vibrate at different tones, to investigate what happens in the ear when tones combine. Helmholtz reckoned that the reason discordance is unpleasant is that sound waves from notes close to each other in pitch stimulate cochlear vibrators that are physically close to each other, causing a kind of discomfort. Helmholtz turned all his musical experiments into a highly successful lecture about how sound moves and is perceived, in which he said that all notes are really chords, of which the human ear hears only the primary note.

At the time there was another moving force which was also exciting great interest: electricity. Nobody knew what it was or how it traveled. Did it move, like sound, in waves? It had been known for some time that electricity had an effect across space. In the

234 99 69
234 184 145   1780s Luigi Galvani[234] had used an electrostatic generator to make a frog's leg twitch at a distance. Joseph Henry had seen how a one-
234 216 179   inch spark magnetized needles thirty feet away. In 1879 David Hughes, a teacher of natural science in Kentucky, had heard a microphone make a sound when sparks were created in a nearby generator. So when Helmholtz offered one of his pupils, Heinrich Hertz, the chance to write a dissertation, Hertz said he wanted to

investigate the mystery of traveling electricity. In 1887, at Karls-
ruhe Technical College, Hertz gave his now-famous lecture-hall
demonstration on how electricity propagates.

First, he generated a large spark between two metal balls. Two
feet away, he set a wire shaped into an incomplete square, with the
broken ends almost touching. When Hertz triggered the large
spark, a tiny spark appeared in the distant wire gap. Hertz used
reflectors of zinc to show that the electricity was moving in waves
that interfered with one another, just as light did. He used prisms
of coal tar pitch to show that the waves were refracted the way
glass refracts light. The waves also went through a wooden door.
Further tests showed that changes in the frequency of the current
causing the spark made waves of different lengths. So electricity *did*
behave like light.

In 1865 Hertz's demonstration would have effects nobody could
possibly have ever foreseen, thanks to the elopement of an Irish
whiskey heiress with her Italian lover. Annie Jamieson, the heiress,
had a good singing voice, but her father disapproved of a musical
career. To distract her ambition, he sent his daughter on a tour of
Italy. During the trip she met and fell in love with another would-
be opera star, called Giuseppe, whom she eventually married. The
couple settled in a village near the northern Italian city of Bologna,
and in 1874 Annie gave birth to a son she named William.

William was crazy about technical things and, as a teenager, was
encouraged by a local physics teacher to start experimenting with
the new traveling electric force. In 1895 he rigged up one of Hertz's
spark-gap generators, linking it to a Morse telegraph key, which
he used to make and break the connection so that the sparks were
generated in sets of three (Morse code[235] for the letter *S*). William
found he could send these interrupted electric waves a kilometer,
and with improved equipment the distance more than doubled.
When the Italian authorities showed no interest in this trick, William went to England (he was bilingual and had attended school
there); after a series of demonstrations for the British Post Office,
he succeeded in transmitting a signal across the English Channel.

On December 12, 1901, using aerials held up by kites, he sent
the Morse code signal for the letter *S* from England to St. John's,
Newfoundland, over thirty-five hundred kilometers away. There-
after, transmissions took place from captive balloons, from ship to

235 30   29
235 114 85
235 275 237

shore, from airplanes and even from the *Titanic* (in time to save over seven hundred survivors). In 1910 the signals were used successfully in the manhunt for the world's most famous criminal, the wife-killer Dr. Crippen. And all the time William Marconi's signals[236] went farther and farther around the Earth, at one point all the way from London to Buenos Aires. So electric waves were not behaving like light after all, since they were clearly being "bent" in some way, following the curvature of the Earth.

236 53 42
236 299 269

In 1902 Oliver Heaviside and A. E. Kennelly remarked that Hertz's electric waves had been reflected by zinc mirrors, and this suggested that there might be a kind of giant reflector in the sky. In 1925 Edward Appleton, an English physicist, used the new BBC radio transmitters to aim a number of signals straight up at the sky, to measure how long it took for them to return and in what ways they had been changed. The speed at which electricity moved (186,000 miles per second) told him that the reflector, whatever it was, operated about sixty miles high. A little later two Americans, Gregory Breit and Merle Tuve, found that signals of certain frequencies were also being reflected from even higher altitudes. Repeating these tests worldwide, they found that the frequencies that were reflected would vary with the time of day, the season and the geographic location.

There was only one thing that would reflect radio waves besides metal reflectors like the ones Hertz had used: ionized atoms, which had lost one or more of their electrons. These atoms became positively charged and would reflect electronic signals (which were negative). This theoretical explanation for signal reflection had already been confirmed in 1910, by the work of a French researcher, Theodore Wulf. Nine hundred feet up, at the top of the Eiffel Tower in Paris, Wulf had shown that the ionization was greater there than it was at the foot of the tower. Then between 1911 and 1912, an Austrian physicist, Victor Hess, took matters considerably further, with a number of balloon flights to an altitude of sixteen thousand feet. The flights revealed that the levels of ionization went on increasing, the higher Hess went. The following year a third intrepid researcher rose to twenty-eight thousand feet and found that levels of ionization there were twelve times what they were at sea level. It looked as if the source of the ionization was some kind of radiation coming from outer space. This radiation became temporarily known as Hess-rays.

Meanwhile, one of Appleton's problems with radio transmissions was that the signals would sometimes inexplicably fade. Fading occurred most often at night and when there were periods of high sunspot activity. Clearly, radiation from the Sun was knocking electrons from the gas atoms in the upper atmosphere and ionizing them. And since the loss of radio signals happened most of all during the high point of the eleven-year cycle of solar activity, it looked as though when the Sun was busy, it bombarded the Earth's atmosphere with increased levels of radiation, causing high levels of radio-wave disruption.

This theory left one rather awkward, unexplained problem. Hess and the others had found that at the highest altitudes, the levels of ionization remained unchanged night and day. So something out there, other than the Sun, was generating constant, ionizing radiation. In 1933 an AT&T engineer, Karl Jansky, who was looking for the source of interference ruining radio reception on the latest luxury liners,[237] discovered that at certain frequencies constant static 237 8 *15* was being caused by some kind of radiation coming from all over the Milky Way. Four years later Grote Reber, an unknown radio repairman in Wheaton, Illinois, built a chicken-wire antenna in his back garden and made the first radio map of the sky. The radiation, he found, was coming from all over the universe. Radio astronomy was born, and "Hess-rays" had now become "cosmic rays."

One of the other things that had been noticed about the eleven-year solar cycle was that it seemed to cause the weather to behave in a similar, cyclical pattern. In the early 1930s a young American physicist, John Mauchly, teaching at Ursinus College, became interested in correlating the data on this behavior. During summer vacations, while he was still a student at Johns Hopkins University in Baltimore, Mauchly had worked at the Weather Bureau and the National Bureau of Standards. There he learned that although weather forecasting data had been collected in the United States for over a hundred years, nothing had been done to analyze them. Mauchly decided that the figures might help to provide information on long-term weather patterns — perhaps making it possible to predict droughts, rainy spells and other such agriculturally ruinous phenomena.

However, it would take too long to process the massive amount of data, unless some faster method were found to do the tedious calculations. In 1934 Mauchly happened to be in Chicago visiting the Barthold Institute, whose director had been a friend of Mauchly's father for some years. There he saw a group of cosmic-ray-particle researchers using vacuum tubes that would react extremely quickly to signal input. These physicists were using the tubes to register up to one hundred thousand cosmic-ray-particle impacts a second. Mauchly realized this technique might be adapted to do the calculations he needed for his weather-data work. But before he could do much about it, World War II would involve him and his ideas in something very far removed from weather forecasting.

The problem with the war effort early on was that it was way too successful. Allied research and development produced different weapons almost every week. These armaments involved new kinds of explosives, guns, aiming devices, propellants and so on. All these new technologies complicated even further the already-complicated business of firing a weapon, which involves a great deal more than just pulling the trigger.

When any weapon fires, many factors interact to influence the accuracy with which the projectile reaches the target. These factors include a truly mind-boggling number of considerations: type of gun, type of shell, type of propellant, type of barrel, rate of propellant burn, type of primer, type of cartridge case, pressure of gases in the barrel, pressure of gases outside the barrel, pressure of back shock, projectile velocity, friction created by the shell leaving the barrel, gun-barrel bore, barrel distortion from the heat of the explosion, density of the air, near-muzzle blast-absorption system, resistance on the shell in flight, forebody drag on the shell, shock waves created by the shell, shell-skin friction, shape of the shell, angle of trajectory, rate of shell spin, shell mass, ambient air temperature, wind direction and speed, humidity, force of gravity, relative height of target, type of target, impact type required, target penetration required and impact angle. And most of these factors were also affected by the terrain in which the gun sat, the movement of the Earth at the time and the position of the Moon!

With these inconveniences in mind, it must be clear that the principal problem involved in firing a gun was not the shot but the arithmetic needed to work out how to do it. The sums came in the form of a small booklet delivered with each weapon. It contained artillery tables that would tell the gunner how the weapon should be fired, under all possible conditions. In the United States,

these tables were being prepared at the Ballistic Research Labs[238]   238   118   90
in Aberdeen, Maryland, by a number of women mathematicians.

Their task was daunting in the extreme. A typical single trajectory required 750 multiplications, and a typical artillery table for
one gun included over three thousand trajectories. Working round
the clock, with the mechanical calculators then available, the mathematicians at Aberdeen took thirty days to complete one table. By
1944 requests would be arriving at the lab for six new tables a day.
Some faster way to do arithmetic had to be found if the Allies were
to go on firing their guns to victorious effect.

In 1942 some of the women table makers started taking courses
at the nearby Moore School of Electrical Engineering in Philadelphia, where John Mauchly was then teaching (one of the mathematicians married him). It occurred to Mauchly that the arithmetic
anxiety might be assuaged with a device that could add, subtract,
multiply and divide, and then store the number until it was needed
for further calculation. Storage was the key to success because his
earlier work on statistics had shown him that most mistakes are
made at the point where a previously calculated number is picked
up again, to be used in the next stage of calculation. So what
Mauchly had in mind would need to work extremely fast. He had
briefly used just such a device in his weather-data calculations. It
was the vacuum tube the cosmic-ray researchers had been using,
which could react to input one hundred thousand times a second.

In 1942 he sent a memo to his military superiors, with the war-
winning title "The Use of High-Speed Vacuum Tubes for Calculating." The Army lost or ignored it. In 1943 the proposal was resub-
mitted and, this time, was accepted. From then on, Mauchly and a
colleague, J. Presper Eckert, worked on the project. The result of
their efforts was a giant machine, switched on for the first time at
the Moore School, in 1946, too late to speed up wartime calculation
of artillery tables. The machine was called the electronic numerical
integrator and calculator (ENIAC). It had cost $800,000; was one
hundred feet long, ten feet high and three feet deep; contained
almost eighteen thousand vacuum tubes; and consumed 174 kilo-
watts of power. The joke at the time was that when ENIAC
switched on, the lights of Philadelphia dimmed.

Because of the tedious and time-consuming way ENIAC had
to be wired up for each job, it was known by its operators as the

"machine from hell." Tedious it may have been, but within a few years ENIAC would affect the lives of everybody on the planet. Meanwhile, the tedium: ENIAC used groups of ten vacuum tubes, each "decade" representing units, tens, hundreds and so on. These decade groups could each be fed with an electronic pulse. In any decade a number of pulses would turn on the same number of vacuum tubes. So four pulses sent into the "units" decade would activate four tubes. Two pulses sent into the "tens" decade would activate two tubes. The number now stored was twenty-four. To add another fifteen, you simply turned on five more tubes in the units decade and one in the tens decade. Then to retrieve the sum, all that was needed was to find out how many pulses it took to return the entire system to zero (by turning all the tubes off). In this case, the relevant pulses required would be three in the tens decade and nine in the units decade, so the sum was thirty-nine. Each pulse took only 0.02 milliseconds.

In spite of how complicated the process sounds, once it was wired up, ENIAC was able to calculate in a day what would have taken one of the Aberdeen mathematicians a year. In honor of the women it replaced, the machine was called by the name used in their job description: "computer." ENIAC's very first task was the one that changed the world. It was used to mathematically model the process of detonating the first hydrogen bomb. The calculations took several months, and the result of the work (thanks ultimately to Jethro Tull's agricultural interests) vaporized the Pacific atoll of Elugelab on November 1, 1952. Plowshares had been turned into swords.

In one of history's fascinating pieces of minutiae, it was not so much the hydrogen bomb that would change the world as the fact that ENIAC's fragile vacuum tubes left much to be desired. . . .

15 New Harmony

S OMETIMES events on the web of change follow a path that eventually comes full circle. One of these paths began with a discovery that perfectly fulfilled technology's aim: to be so pervasive and user-friendly as to be taken for granted. It all began with the vacuum tube.

In the 1940s, one of the most frightening things for crews flying B-29s during World War II was not the enemy flak exploding all round them, but the tiny vacuum tubes responsible for keeping their planes in the air and on course. The B-29 had over a thousand such tubes (as did many other war machines at the time); if just one of them failed, the effect could spoil everybody's day. The problem was that the tubes operated everything electrical, acting as switches and amplifiers in heaters, instruments, radios and engines.

The vacuum tube worked by harnessing the stream of electrons shooting from the hot filament[239] to the tube's metal baseplate. This electron stream was used to boost very weak electric charges (either radio signals or output from batteries) so that they would be strong enough to do their job. Unfortunately, the glass vacuum tubes broke easily, the filaments corroded or snapped, and the vacuum could leak away. Any one of these events would cause a malfunction in the tube and then in the switch it operated and then in the engine (or whatever else that switch controlled). One last imperfection in the vacuum tube was that it took several seconds to warm up, which was less than desirable if the tube happened to be the amplifier in a detector that would warn a ship of the

239 52 31

imminent approach of a torpedo. Not unnaturally, people were keen to find an alternative that would be more reliable, longer-lasting and faster.

Something solid, rather than an easily breakable glass bulb, would be ideal. For some years before the war, Bell Telephone Laboratories had been looking at the way materials called semiconductors behaved (with the vague hope of improving the telephone system), but with the advent of the war, this work was shelved. In 1945 Bell picked up where it had left off and hired a team of researchers, working under W. B. Shockley.

This was the group that made the breakthrough for which, in 1956, they would be awarded the Nobel Prize for physics. They found that if certain impurities are added to the atoms in certain semiconductor crystals, the atoms either take on extra electrons or, depending on the impurities, lose electrons, leaving "holes" in their place. Putting such a crystal in an electrical field causes the negative electrons to jump to a positive electrode, or causes the holes to do the same thing. Either way, the result is a unidirectional current that can be used like a vacuum-tube particle stream, to amplify a charge or to act as a switch, because the effect can be reproduced as fast as the field acting on the crystal can be switched on and off. This can happen thousands of times a second. The new device, called a transistor, does the work of a vacuum tube. Since it consists principally of a solid piece of crystal, it is also much less fragile than a vacuum tube, lasts longer and works much faster. Shockley developed the transistor using a semiconductor material called germanium. Later, because the extremely rare germanium was worth its weight in gold, he switched to silicon, the material common in all switching systems today. Initially, however, the germanium being used was recovered from zinc ores found in, among other places, the tristate area of Missouri, Oklahoma and Kansas.

Germanium had originally been discovered in 1886, not surprisingly (given the name) by a German professor of analytical and technical chemistry at the Freiberg School of Mining. His name was Clemens Winkler, and he was an analytical genius. So when a large vein of the rare mineral argyrodite was discovered, he was asked to assay it. He found the sample to be 74.7 percent silver, 17.1 percent sulfur, 0.66 percent iron oxide, 0.22 percent zinc oxide and 0.31 percent mercury. So what was the other 7.01 percent? After much chemical investigation, Winkler announced a new mineral which he named germanium (the material Shockley would use) and went back to his other main interests. The first of these was the development of his gas burette, one of the forerunners of the modern technology that guarantees a clean environment in the vicinity of such places as metalworks and factories.

The burette helped to analyze the contents of the fumes[240] in a chimney by measuring how much gas was absorbed by a liquid, which reacted to the gas by changing color according to how much gas it had absorbed. This information was of interest to industrialists because it saved them money, especially in operating blast furnaces. These worked by using carbon monoxide and solid carbon to reduce iron oxides. So anything that could identify the presence of unused carbon monoxide would enable the factory to recycle it. Gas analysis also quantified the amounts of unlit fuel gas and other potentially recyclable fumes and led to the more efficient redesign of furnaces and chimneys in general. This kind of analysis would also reveal the presence of poisonous or explosive gases in underground mines. It may be for all these reasons that Winkler is known as "the father of technical gas analysis."

Fortunately for this story, the zigzags of history lead rather quickly away from this somewhat underwhelming aspect of chemistry, because Winkler's primary interest lay in cobalt. His father managed a large cobalt works, and Winkler had spent his holidays there as a boy and went on to work in the trade. At the time, cobalt[241] blue was widely used in the dyeing industry and even more widely in making pottery and enamel. A particularly pure form had been developed by a chemist in Birmingham, England, and his cobalt was much in demand in the Sèvres factory near Paris. In the eighteenth century Wedgwood[242] had already made blue a very popular color, with his blue-and-white neoclassical crockery designs. Wedgwood had begun his career repairing Dutch delftware, itself an imitation of the very expensive and outrageously fashionable Chinese porcelain.[243] Porcelain had been arriving in Europe since the early seventeenth century; when it first began to be exported from China, the fashion there had been for blue-and-white designs.

Cobalt had first been used by the Chinese during the Ming dynasty, when some of the greatest porcelain in history was made.

240 110 *76*
240 197 *157*

241 177 *139*

242 124 *94*

243 11 *18*
243 291 *259*

But ceramic skills had been developing in China for centuries, since the invention of the glazing process in the second millennium B.C. The industry benefited from the ancient Chinese tradition of giving ceramics as tribute to heads of government. On some occasions, as many as fifty thousand pieces would be taken to the imperial court. By the fifteenth century the Ming capital was at Nanking, near the great porcelain-making factory of Ching-te-Chen, where the first blue-and-white pieces were made as novelties for the court. Cobalt was what made possible the famous Ming-vase blue, because it could be painted directly onto the vessel body before the glaze was applied and then would not run under the very high temperatures at which porcelain is fired. This stability under fire meant that artists could be ambitious with their designs, which included paneled scenes of real and mythical animals, birds, children, fishing boats, pagodas, bridges and gardens, as well as Taoist and Buddhist symbols and Chinese characters.

The Chinese had originally come across cobalt as their mer-
244 90 65 chants traveled along the Silk Road,[244] the great trading highway running from China through Central Asia to Syria. The route had originally been opened under the Han dynasty, in the third century B.C., in order to trade with the Roman Empire. By the early Middle Ages each nation along the road took care of its maintenance, garrisoning the roads, protecting the caravans and collecting tolls. There were well-known checkpoints where, at various times through history, Greek, Arab, Roman, Indian and Persian traders exchanged goods with the nomad merchants who would then travel all the way back to China and pass the merchandise on to their Chinese clients. Nomads were used because Chinese citizens were not permitted to travel abroad and risk contamination from "barbarians."

It was a Persian link which encouraged Ming porcelain makers to use cobalt, because it was in Persia that the Chinese indirectly discovered the mineral (and named it Muhammadan blue). This discovery probably happened in the city of Kashan, where there were local mines from which the town potters obtained cobalt ore. The earliest known blue Kashan ceramic bowl and tile dates from 1203, just before the tile-making family of Hassan ibn Arabshah, who flourished for more than a century, created the beautiful decorated pottery which helped spread Kashan's reputation throughout Persia. It was probably the Kashan potters (who had found that cobalt tended to run when used with earthenware and then successfully tried the color on samples of Chinese porcelain) that gave the Chinese the idea of using the dye in the first place.

A characteristic technique of Kashan tiling, which was especially prized for use in mosques, was the way in which the blue was used to pick out in relief Arabic inscriptions of Persian verses. Other styles involved groups of tiles cut up into stars, hexagons or crosses and tiles formed in large, decorative panels. Many mosques and palaces were also decorated with distinctive and complex patterns of very small pieces of tile, each one of which carried a part of the overall design. This technique can still be seen on the Esfahan mosque of Maidan-i Shah, and similar techniques appear in the Alhambra, in Granada, Spain.

The Islamic ceramicists had originally learned their sophisticated techniques from Greek mosaic masters, who spread out from Byzantium in the early Christian period to centers of production in Syrio-Palestine (particularly the great school of Antioch) and Egypt, to train others to meet the increasing demand for their art. Although a form of mosaic found in Mesopotamia dates as far back as 5000 B.C., and there was a long tradition of production stretching through Greek times to the late Roman Empire, the craft reached its zenith in the fourth and fifth centuries, after Constantine had made Christianity the official state religion of Byzantium, and the Christians were then free to come out of hiding and begin to organize themselves.

They probably chose mosaics with which to create the monumental and dazzling decoration of the churches that soon sprang up all over the Byzantine Empire because the Christians had spent years literally underground, as a proscribed movement. Mosaics were particularly well suited to the dimness and uneven surfaces of subterranean caves because they reflect so much light. The surface to be decorated was first scratched and then covered with several inches of cement to give an even ground. Then the small glass tesserae were squeezed into the still-moist cement surface. The tessera pieces were usually rectangular or square, rarely larger than three-quarters of an inch across. Each glass piece was colored with metallic oxides. Gold and silver foil were also sandwiched between the glass and an enamel base, which was often colored red to give the gold a richer effect. So that they would reflect the maximum light, the tesserae were tilted at an angle of 30 degrees.

Mosaic colors were relatively unsubtle and their real effect could be appreciated only from a distance and when used on a grand scale. Then, the effect was stunning. Some of the greatest of all mosaic work still exists at Ravenna, seat of the Byzantine viceroy of Italy, where there are extraordinary fifth- and sixth-century examples. The most beautiful are probably in the tomb of Galla Placidia — decorated with religious scenes rich with vegetation, animals and birds — considered perhaps the most beautiful mosaic in early Christian art. The tomb stands next to the church of San Vitale, which contains a marvelous example of mosaics at their monumental best, recording the consecration of the church in the presence of the emperor Justinian and his empress Theodora, in full ceremonial regalia.

One of the most politically interesting mosaics visible today stands in Rome, near the Scala Sancta in the Lateran square. Here an eighteenth-century open apse displays a mosaic, carefully restored three times over a period of one hundred years and finally placed there by Pope Benedict XIV, in 1743. The mosaic dates from the ninth century, and it has received so much papal care and attention because of its subject matter. On one arch the mosaic shows Christ enthroned, giving the keys to Saint Peter and a banner to the emperor Constantine to commemorate the emperor's recognition of Christianity. On another arch the scene is of Saint Peter handing the pallium to Pope Leo III and a banner to Charlemagne. This scene memorializes the agreement between Leo and Charlemagne, at the latter's coronation on Christmas Day, A.D. 800, formalizing the pope's spiritual supremacy over all secular authority.

Both mosaics implicitly refer to the document which originally gave popes these special powers and which was known as the "Donation of Constantine." The Donation took the form of a three-thousand-word document addressed to Pope Sylvester I and signed by Emperor Constantine. Legend has it that the emperor caught leprosy, which then miraculously vanished when he embraced Christianity. In gratitude for this miracle, he made the "Donation." But whatever the reason, the Donation gave the popes unprecedented powers and primacy over the universal church. It also elevated the Roman clergy to the nobility and conferred on Rome the provinces and cities of Italy, Lombardy, Venetia and Istria. When all was said and done, because of the Donation the pope was now in charge of the entire Western Empire. So it was his right to crown Charlemagne as his spiritual vassal.

Not surprisingly, the Donation became the church's fundamental instrument of power for centuries, cited by no fewer than ten popes and taken as a document of precedent by all medieval lawyers and theologians. And all this happened in spite of the fact that the Donation was a clumsy fake. It was probably written by one of the pope's advisers around 750, when Rome was cutting loose from Byzantium, moving closer to the increasingly powerful Frankish kings and needing to establish clear authority over them.

The forgery might never have come to light had it not been made public, in 1435, when the pope picked the wrong man with whom to have a quarrel. The man in question was the Spaniard Alfonso V, who was king of Naples because, he said, the recently dead Queen Giovanna had adopted him. The pope contested this claim, and the two men were soon bitter enemies. Alfonso looked around for ammunition to use against the papacy, and whom should his eye light on but his secretary, a controversial young prelate named Lorenzo Valla.[245] He had previously been professor of rheto-   245 302 273
ric at the University of Padua (the MIT of the period) and, at the age of twenty-nine, was one of the leading humanists in Italy.

Valla had extensively studied classical Latin, using the newly found ancient manuscripts revolutionizing everybody's ideas of the past. He was Europe's leading expert in Latin style and in 1444 wrote a book on the subject that became the definitive work for another three hundred years. So Valla was in a good position to delve into records of the murky papal past, in search of propaganda for his royal boss.

Neither of them could have remotely expected what Valla would find. In examining the Donation for linguistic quality, he found it riddled with bloopers. To begin with, the Latin language in which it was written was "barbaric" and certainly from a period much later than Constantine's era. The Donation also referred to Constantinople as a "patriarchal see" (which at the time hadn't yet been established) and spoke of Constantine's "golden diadem" (when in fact the emperor would have worn a head-covering of cloth).

Valla's work was a typical example of what the humanists had learned from their examination of ancient manuscripts. The way to understand what classical authors had originally meant was to examine their writing critically, using grammatical and syntactic analysis, allied with historical analysis and common sense. These techniques allowed scholars to collate and synthesize various corrupt examples of a text and to produce a definitive version. It was a literary exercise that would eventually be extended to cover classical scientific texts; consequently, it laid the groundwork for the Scientific Revolution over a century later. But for the moment, its use had put Valla in deadly danger. It was unthinkable to do what he had done — and expect to get away with it. The fact that he had, and did, was because he had a very powerful friend in a very high place.

Nicholas of Cusa was one of the major intellects of the church. Like Valla, he had attended the University of Padua, where he had taken a degree in canon law. Ordained in 1436 at the age of thirty-five, he was papal envoy by thirty-seven, a cardinal at the exceptionally young age of forty-five and governor of Rome at fifty-eight. As one of the church's leading lawyers, he had argued the papal position at the Council of Basel and, in 1434, wrote a major work in which he, too, had discredited the Donation of Constantine. But unlike Valla, he had done so primarily to show that the Holy Roman Emperor was not dependent on the pope for his authority. This formed part of his general argument that popes ought to defer more to the opinion of church councils. So Cusa was sympathetic to Valla's case and was probably instrumental in keeping him alive after word got out of his discovery of the forgery. And although the church suppressed it, the news would eventually add to the many criticisms of Rome that would lead to the Protestant breakaway a century later.

Meanwhile, Valla's protector, Nicholas of Cusa, was a man very much ahead of his time. Well before Copernicus,[246] he proposed that the Earth was a sphere, spinning daily on its axis and orbiting the Sun; that heavenly bodies were made of the same material as the Earth; and that the apparent circular movement of the sky around the Earth was a relativistic observation. Long before Galileo,[247] he discussed dropping things from towers to find out why they fell as they did. And well before anyone else, he described rules for experimentation, took the pulse, forecast the weather, argued for an infi-

246 128 97

247 130 97
247 158 *126*

nite universe, invented reading glasses and possibly supplied the data for a new map of the world.

It may have been this last geographical interest of Cusa's that linked him to Paolo Toscanelli, a fellow student at Padua, who was to become one of the period's greatest mathematicians and geographers and who sent Columbus the map suggesting that he could reach Japan by going west across the Atlantic. Toscanelli and Cusa were lifelong friends, and Cusa dedicated one of his treatises on geometry to him. So in 1464, when Cusa died in the small Italian town of Todi, Toscanelli was there as a witness to his will. So, too, was the man who takes this circular tale to its next stage.

He was Fernao Martins de Roriz, a canon from Lisbon, with whom Toscanelli had exchanged correspondence ever since the Italian had conducted a series of interviews about various aspects of geography from the well-traveled Portuguese delegates at the Council of Florence,[248] in 1459. At that time, the Portuguese were already the West's greatest explorers, having discovered Madeira, the Azores and the Cape Verde islands and being within only six years of crossing the equator, as they sailed down Africa in search of a route to the East. The Portuguese purpose was to get their spices directly from the suppliers instead of having to pay the Turkish middlemen operating in Istanbul[249] since 1453, when the Turks had taken the city and changed its name from Constantinople.

The extraordinary success of the great Portuguese explorers was based on their jealously guarded secret of how to run up and down the latitudes.[250] This involved measuring the angle of the Pole Star with a quadrant (a Portuguese innovation) in order to determine how far north or south they were. The farther north they were, the higher the star stood in the sky. Using this technique, they could be sure of getting home on a return journey. They just headed north until the Pole Star stood at 38 degrees latitude, turned east and sooner or later sailed straight into Lisbon (which is located at almost exactly 38°N).

Martins de Roriz was secretary to the Portuguese special commission on navigation, set up by the court to organize seaborne exploration. By 1474 Portuguese ships had reached Greenland, and Europe was rife with rumors that Portugal was also conducting secret, experimental voyages far out into the Atlantic to test new navigational techniques. Perhaps the most important, and most

248 301 273

249 91 65

250 265 224

closely guarded, secret was finding latitude by measuring the Sun's position, so as to be able to run the latitudes by day, after crossing the equator going south and losing sight of the Pole Star.

Going for the African route to the East was an unbelievably risky thing to attempt because it involved leaving sight of land (nobody had ever done that before) and taking the most direct route for the Horn of Africa by sailing straight down the Atlantic, instead of hugging the African coastline. The Portuguese were encouraged in this dangerous venture by news brought back from a secret overland mission to the East by Pero da Covilhã. He was fluent in Arabic and had already spied for Portugal in Morocco. On his return from the mission in 1490, Covilhã reported that pepper and ginger were grown in India and that cloves and cinnamon were brought there from islands farther east. He had also found out the dates on which Arab ships typically left Indian Ocean ports to make the best use of the monsoon winds.

Thanks to Covilhã, the Portuguese already knew where to go for their supplies once Bartolomeu Dias rounded the Cape of Good Hope in 1487. Ten years later, Vasco da Gama began his epic voyage to India, spurred on by news of Columbus's voyage. Da Gama first sailed deep into the mid-Atlantic and then due south about 30 degrees, before turning east and finding that he was only one degree north of the Cape. Using the sun to run the latitudes had worked.

Less than six months after da Gama's return, in 1500 Pedro Álvar Cabral was leading a fleet of thirteen ships on the same route when he was blown farther west than he intended. On April 23 he discovered Brazil.[251] Another ship immediately followed the same route and planted the flag. Then in 1501 a third fleet set sail for Brazil, commanded by the Italian Amerigo Vespucci, who would give his name to the New World. A few years later, the German cartographer Martin Waldseemüller called the new continent (after Vespucci's first name) America.

Even though the first reports coming back from Brazil gave the distinct impression that all the country had to offer was parrots, monkeys, naked savages and dyewoods, there was something else to recommend it. There was no Inquisition in Brazil. So it became the first of many New World countries where people went to escape from discrimination. In this case, the refugees were Portuguese Jews (or rather, ex-Jews, because they'd all been forced under threat

251 94 66

of torture to become Christians). The Portuguese authorities called them New Christians, either deporting them or making it difficult for them to find work.

So when some New Christians asked to be given a chance to emigrate to Brazil, the crown had a perfect opportunity to get rid of them and (with one eye on an expansionist Spain) strengthen its presence in Brazil. In 1502 a monopoly concession was granted to a New Christian consortium headed by Fernando de Noronha, on condition that the consortium undertook to send back no fewer than six ships a year, to explore three hundred leagues of new land every twelve months and to build and maintain fortifications. The New Christians became Brazil's first settlers and, around Pernambuco, the first to run sugar plantations.

Back then, sugar production wasn't the uninteresting backwoods agribusiness some people might consider it today. In the seventeenth century, sugar was as valuable a commodity as petroleum is now, and with the new Brazilian plantations, Portugal was able to dominate the sugar market. This was like having a license to print money. In 1630 the Dutch occupied Pernambuco and then all of northeast Brazil, and life for the New Christians suddenly got a whole lot better. The Dutch believed that nobody should be persecuted for his or her religious beliefs, and in this tolerant atmosphere New Christians were soon revealing that they had never really renounced their Jewish religion at all.

Their newfound freedom of worship meant so much to them that when the Portuguese attempted to reconquer the territory, the Jews fought for their new Dutch friends. When the Dutch lost and Pernambuco capitulated, large numbers of Brazilian Jews left for New Amsterdam (now New York), where they set up the first synagogue in North America. Others went to the other Amsterdam, where they could also be assured of safety from persecution. One of these refugees to Europe was a young boy, Andrade Velosinho, who later in life was to become the powerful opponent of another Portuguese Jew, called Baruch Spinoza, whose parents had fled Portugal and settled in Amsterdam.

Spinoza was one of the few people ever to achieve the doubtful distinction of being declared a heretic by both Catholics and Protestants. In 1656 his unorthodox ideas also got him excommunicated by the Jewish community. Spinoza was a quiet, scholarly man who spoke Portuguese, German, Dutch, Latin and Hebrew and who had at one point started training for a rabbinical career. Being short of money, he took up a career as a lens grinder and in the course of time met and advised both Christian Huygens[252] and Gottfried Leibniz,[253] two of the century's major scientific thinkers.

252 84 56
252 132 98
253 82 55
253 231 194
253 309 279

Meanwhile, his philosophical writings were getting him into trouble with the established Christian churches. In 1670 he anonymously published a treatise on theology, calling for complete freedom of speech and thought. Even in tolerant Holland, this proposal was revolutionary. Spinoza also argued that miracles were seen only by those who wanted to see them; that if God had laws, they were inherent in natural laws; that there was no afterlife; that knowledge of nature was knowledge of God; and that the only rational language with which to describe the universe was the language of mathematics.

It was probably Huygens who drew the attention of the English Royal Society to Spinoza's mathematical work. In 1661 he began to correspond with Henry Oldenburg, who was about to become the society's first secretary and who paid Huygens a visit later that year. In the last quarter of the seventeenth century, the society became interested in Dutch scientific research, particularly (through Spinoza) in the work of Anton van Leeuwenhoek,[254] who ground lenses for microscopes. In 1676 Leeuwenhoek caused a sensation at the society when he sent sketches of protozoa and of rupturing cells releasing their protoplasm. This was the first time the microscope had been used to show minute living organisms, and it changed the scientific world at a stroke. If there were, as Leeuwenhoek's work seemed to suggest, living creatures too small to see, what else was there to be discovered?

254 83 55
254 310 279

Throughout the eighteenth century, it must have seemed that hardly a day went by without some amazing revelation from science and technology. Almost every month came discoveries relating to gravity, electricity, chemistry, respiration, photosynthesis, geology, physiology and biology. Above all, by 1813, the extraordinary new machines driven by steam power were radically changing the world. This was the year in which the mysteries of science attracted the attention of one particular group of social misfits, vacationing on the shores of Lake Geneva. They were the Romantic poets Byron (with his mistress) and Shelley (with his child bride, Byron's mistress's stepsister).

Photographic illustrations from Lombroso's *Criminal Man* published in 1876 when he was professor of forensic medicine in Turin. The obsession with skull shape which characterized Lombroso's work formed part of the general mania for anthropometry and included a German classification system requiring 5,000 measurements from one head.

HEAD OF CRIMINAL

HEAD OF CRIMINAL

The great triangulation-method Ordnance Survey of India, completed in 1876. The first triangle was measured from Madras, whose longitude was determined in 1807 by star fixes to be exactly 180° 14' 20" east. This first triangle formed the base for the next, and so on. During the course of the survey Mount Everest was found and named (after the survey chief); its summit measured as 29,002 feet above sea level.

INDEX CHART
TO THE
GREAT TRIGONOMETRICAL SURVEY
OF
INDIA

In the *Illustrated London News* of 1912 (the year a radio message saved 710 of the *Titanic's* survivors), an artist's impression presents Guglielmo Marconi as an already famous figure. The drawing shows early experiments, when Marconi was still using metal-sheet aerials over a very short distance.

A detail from the great sixth-century mosaic depicting the Emperor Justinian and his family, in the church of San Vitale, at Ravenna, Italy. We see the Empress Theodora with her ladies-in-waiting. Ravenna was the seat of the Byzantine governor of the west, and the rich mosaics in its churches are among the finest in the world.

The film monster (played by Boris Karloff) created by Mary Shelley's young hero, Victor Frankenstein, in the first of a new genre now known as science fiction. It is likely that the author gained the scientific knowledge on which she based her story from Sir Humphry Davy's famous pamphlet on chemistry, in which Davy talked about "discovering nature's hidden operations."

A contemporary view of the battle between the English fleet (whose flags carry the red-on-white cross of Saint George) and the Spanish Armada. Typical of the old-fashioned Spanish fleet is the lumbering galley seen side-on in the foreground, whose banks of oars severely limited the number of guns on board.

The subscription room at Lloyd's of London, where insurers who accepted a part of any risk wrote their names and the amount of cover they would provide on the side of the policy document. This international insurance business was only made possible by the development of reliable international law to regulate disputes between parties of different nationalities.

A Gillette safety razor, pictured in Paris only one year after Gillette had set up his first factory in Boston. The razor cost two days' pay ($5.00) for the average worker and twelve blades (giving a total of thirty shaves) cost $1.00. In his first year Gillette sold 51 razors. In his second year he sold 90,844.

An 1879 print showing Smeaton's Eddystone lighthouse, finally being replaced (by the structure being built in the foreground) after 120 years. Smeaton had personally screwed in place the gilded ball on top of the cupola. The first light was provided by 25 six-pound tallow candles and could be seen five miles away. Construction of the replacement lighthouse was completed in 1882.

A classic example of Wedgwood's neoclassical style of pottery. Wedgwood was the first to implement "factory"-style mass production in his pottery; at his new showroom in London, with its new plate-glass windows, he provided the first sales catalogue for patrons to study before ordering their vases, medallions, ornaments or crockery sets.

A contemporary painting of Napoleon's troops resting at Syene, up the Nile, in 1799, after their leader had gone on to Syria. While Napoleon's commission cataloged the antiquities, the ordinary soldiery also left his mark behind, with graffiti that can still be seen in Egypt today. Here, a soldier scratches on a monument the distance to Paris: 1,167 miles.

ELEMENTS

Symbol	Element	w.	Symbol	Element	w.
⊙	Hydrogen.	1	⊕	Strontian	46
◐	Azote	5	⊗	Barytes	68
●	Carbon	5	Ⓘ	Iron	50
○	Oxygen	7	Ⓩ	Zinc	56
⊗	Phosphorus	9	Ⓒ	Copper	56
⊕	Sulphur	13	Ⓛ	Lead	90
◑	Magnesia	20	Ⓢ	Silver	190
⊖	Lime	24	Ⓓ	Gold	190
◑	Soda	28	Ⓟ	Platina	190
◓	Potash	42	⊛	Mercury	167

John Dalton's first table of twenty atomic weights, in which each element's weight is a multiple of the weight of hydrogen. Dalton's concept of indivisible atoms (which could group together to make compounds) revolutionized the science of chemistry, although a better symbology to express these groupings had to await the work of Berzelius.

One night, in Byron's villa, when the after-dinner conversation drifted around to the latest hi-tech gadgetry, somebody mentioned that Erasmus Darwin had apparently galvanized a piece of pasta, making it come alive. The idea of vivified vermicelli set the group to talking about life and death and ghosts and science, and Shelley's wife, Mary,[255] announced her intention of writing a novel about how science and industry were destroying people's lives and how people like the Italian experimenter Alessandro Volta were meddling with forces they did not understand.

255   160   *127*

She wrote about a young scientist called Victor, obsessed with penetrating the secrets of nature and using microscopes and crucibles to investigate the world of the unknown. Victor's great plan is to use the unknown powers of physiology and electricity to create an artificial human being. After spine-chilling experimentation in his dungeon laboratory, Victor comes up with the 1813 equivalent of genetic engineering, which, of course, goes horribly wrong, with terrible consequences. Although the original title of Mary's novel was *The Demon of Switzerland,* it would later become better known by the family name of her young experimenter-hero: Frankenstein.

Mary got many of her ideas about the evil effects of industrialization and technology from Shelley. The Romantics[256] tended to share an antagonism toward materialism and the rise of the middle class that accompanied the growth of industrial manufacturing. Mary's left-wing husband believed that machines enslaved people, removing them from the natural freedom they had earlier enjoyed in the simple, rural village environment (before they had been debauched by life in the town). Shelley got most of these ideas from Mary's father, William Godwin, who knew Coleridge well, was much admired by Wordsworth and was Shelley's idol.

256   232   *195*

Shelley had read Godwin's *Enquiry Concerning Political Justice* while at Eton and in 1812 was thrilled to meet him (and his daughter). Godwin was the founder of socialist thought and was driven by a belief in human perfectibility. He wrote that society should be a voluntary association of autonomous, free and self-reliant individuals. All property should be owned by the community, and nobody should receive more than he or she needs nor be required to give more than he or she is able. The ultimate social idea would be a society of small, self-governing communities.

These thoughts chimed perfectly with the ideas of Robert

257 7 *14* Owen,[257] another of Godwin's admirers, whom he met in 1813. Owen had been the manager of a mill at New Lanark, in Scotland, for thirteen years. There he had instituted changes which, for the period, were little short of subversive. Since most of the mill workers were children, the industrial complex at New Lanark included a school, where, following Owen's beliefs, there were no punishments or rewards. In the factory dormitories, bedding was changed once a week. The factory shop sold goods at cost. There was a sick fund to which workers paid one-sixtieth of their wages. Some conservative writers of the day were shocked at these radical practices; Owen's social experiment became so well known that the factory had thousands of visitors.

For years Owen, who was heavily involved with the cooperative movement, wrote and spoke about his socialist ideals taking the form of a model community based on his experiments at New Lanark. In 1824, he had the opportunity to buy a small community in Indiana, and jumped at the chance. It was situated on the banks of the Wabash River and had been founded by another Utopian, George Rapp, who then decided that life for his flock was getting too easy and planned to move them elsewhere. So the land and buildings were available for Owen to buy. Owen called his venture New Harmony (Rapp had named his community Harmony) and persuaded a group of believers to settle there. Unfortunately, because of financial difficulties, alcoholism and class differences, the settlement was a disaster almost from the start. But for the first four years of its existence, New Harmony was the radical chic center of the United States, with the country's first kindergarten, public library and women's club.

One of the reasons Owen had been able to set up the community in the first place was the financial help he got from William

258 288 *253* Maclure,[258] a rich, retired businessman. Maclure was a liberal thinker whose one passion, apart from philanthropy, was geology. Since 1808, when he had returned from Europe, Maclure had crisscrossed every state in the Union with his hammer and sack, collecting samples and taking notes. In 1809 he published the first geolog-

259 167 *131* ical survey[259] of the United States in the *Transactions of the American Philosophical Society*. It would be the only work of its kind for the next twenty-five years. Attached to the text was a colored map, in which Maclure identified the country's major geological formations.

One of them, in the tristate area of Kansas, Oklahoma and Missouri, noted the deposits from which, one day, Shockley's germanium would come.

Maclure was following a great tradition of physical geography that had originally been established in Renaissance Germany. . . .

16 Whodunit?

SOMETIMES the path of history takes a number of unexpected turns because now and again individuals do something unexpected that causes change to happen in the most unforeseen way. Take, for instance, the unusual individual who triggered a series of events that would, in the end, make a billiard ball every detective's favorite exhibit in the Criminal Hall of Fame. Thanks to that ball, when the modern detective finally identifies the perpetrator, the prosecution has incontrovertible evidence to prove he or she's the one whodunit. All because of the individual who kicks off this story.

He was one of that rare breed often referred to as Renaissance men, those sixteenth-century European intellectuals who knew everything there was to know. Georg Bauer was his name, and he was a precocious young German who began a checkered career by teaching Greek, then Latin and then grammar. Next, he decided to move into physics, then chemistry, then linguistics and then publishing — before he finally settled on a permanent career in medicine. So, like everybody else at the time who wanted the best medical degree in Europe, he enrolled at the northern Italian University of Bologna, where the medical school was jammed with students because it featured the latest techniques in anatomy.

In 1526, after Bauer passed the degree exams summa cum laude, he set off to practice medicine in the mining town of Joachimsthal, at the center of the most productive metal-mining area of central Europe (in what today would be the Czech Republic). Joachimsthal was the site of the great 1516 silver strike. The precious metal was found in such enormous quantities that it produced the longest-lasting coinage in European history. The money took its name from the valley of Joachimsthal. *Thal* means "valley," so the coin was called a *Joachimstaler*. In time this name became abbreviated to thaler. The thalers coined under the eighteenth-century Austro-Hungarian empress Maria Theresa were still being used as payment during World War II. The thaler made a different mark on history, too. It is the word from which the modern "dollar" comes.

Back to our hero: in 1528 Bauer began to write a giant book on everything you needed to know about mining (George was, after all, a Renaissance man and also wrote a book on medicine, as well as the first text on physical geography[260]). The mining text, published posthumously in 1556 under Bauer's pen name of Agricola, was called *De re metallica* (*About Metal*) and became every miner's bible for two hundred years. Bauer covered all aspects of the job, from draining to smelting, assaying, ore crushing, tunneling, ventilating, fossils and the rest. Copiously illustrated with woodcuts, *De re metallica* contained everything that anybody interested in buying a mine needed to know. So if an investor bought a mine that Bauer's assaying techniques indicated would turn out to be full of really valuable minerals like copper, then he could be assured in advance that things were going to go extremely well financially.

260 211 *171*

Apart from digging for precious metal, the mine owner's other way to get rich quick was to put copper and a little tin together and make a bronze cannon: the ideal gift for the prince who had everything. The European sixteenth century bustled with young nations, ambitious kings, shiploads of world explorers and armies fighting the pope. Their primary requirement was firepower. Hardly a week went by without a war somewhere. And cannons were the terror weapon that would ensure success in war because whole cities would surrender merely at the rumor of a cannon on its way. So for the belligerently inclined, owing to the unique talents of Georg Bauer, making war was now that little bit easier.

Making the guns from bronze was easy to do thanks to the church. Prior to becoming involved in casting ordnance, most cannon makers had been bell founders, using production techniques easily adapted to gun-barrel manufacture. The best sixteenth-century cannons came from Germany and Flanders, and each gun cost an absolute fortune. Unfortunately for the cannon makers, however, making a quick profit from their trade wasn't too easy, because their royal clients were always short of money.

The problem was basically one of the royal cash flow. Citizens usually paid their taxes to the local ruler in goods and services rather than in cash, which in any case (in spite of silver strikes in central Europe) was usually in short supply. At a time when the new capitalists in the increasingly commercial cities were flexing their economic muscle, and the old feudal ties were weakening, kings didn't usually have enough ready sources of money. Unfortunately, cash was just what they had to have if they were to build an army and fire their cannons in a war, because the only soldiery at the time was mercenary. And mercenaries always wanted cash on the spot (which was not usually the spot where their employer happened to have his source of revenue). As one of the most famous mercenaries of the period, Gian Giacomo di Trivulzio, had once said to the French Louis XII, "Majesty, [to go to war] ... three things must be ready: money, money, and once again money."

So princes and kings borrowed cash (at exorbitant interest rates) from the new capitalists, many of whom, interestingly enough, had begun their own careers in mining — perhaps the earliest form of capitalism. Because of the considerable amount of expensive equipment needed to set up and operate a mine, the only way to get the venture going was to find backers who would provide financial aid in return for shares of the profit. As the amount of coin circulating in the late Middle Ages gradually increased, these investors were usually the first to get their hands on cash. And as their investments became more successful, some of them made a full-time occupation of investing in enterprises in return for shares.

One such group of financiers was the unique Fugger family of Augsburg, Germany. The Fuggers had been merchants since the fourteenth century, trading in everything from pepper to textiles. By the 1490s one of the sons, Jakob, had moved into the investment business because he had made so much money that he had enough to lend to kings and princes. In return, these royal borrowers either hocked their silver and copper mines to Jakob, pledged their tax income for a number of years or mortgaged whole provinces to him — or all three.

The Fugger family became internationally famous by lending to cardinals who couldn't pay the pope for their hat or to would-be emperors who couldn't pay for their election to the job. The particular would-be emperor in question was Charles V[261] of Spain, and his electors were princes whose votes would go to the highest bidder. In this case Charles's rival for the job of Holy Roman Emperor was the French king, Francis. In 1519, with the aid of massive Fugger loans (distributed mainly to the king of Bohemia, as well as to the princes of the Palatinate, Brandenburg and Saxony and to the archbishops of Cologne and Mainz), Charles pulled off a victory. But the effort had cost him the tax income from Castile, Naples, Sicily, Holland and Austria, as well as the gold and silver which were beginning to pour into Spain from South America.

261   201   *161*

Charles's finances never recovered from the impact of his debt to the Fuggers. In 1556, when he abdicated and retired to the Spanish monastery of Yuste (to devote the rest of his life to his obsession with clocks and watches), he was still broke, with not even enough cash to pay off his servants. Two years later he died of malaria, leaving his son Philip II heir to a bankrupt empire and facing secessionists in Holland who were being given aid and comfort by the commercially up-and-coming Elizabeth of England. By 1586 Elizabeth had signed a treaty with the Dutch rebels, and Sir Francis Drake[262] was sacking Santa Domingo and Cartagena, two of Philip's Caribbean ports where Inca treasure[263] was shipped back to Spain.

262   92   65
263   202   *163*

Philip decided on desperate measures to solve his financial problems. The idea was to invade England with a fully armed massive fleet — for this reason called an *armada* (meaning "armed"). This decision was to prove a foolish mistake. To start with, the venture was very badly organized. Nobody had ever attempted a battle at sea on such a scale before, so Philip had horrendous logistical problems. Worst of all, the English fleet was a formidable adversary. Queen Elizabeth had left the business of maritime warfare to privateering entrepreneurs, most of whom, like Francis Drake and John Hawkins, indulged in thinly disguised piracy. For this reason, the privateers had designed and built for themselves a particular type of ship, good at the cut-and-run action piracy required. Pirate ships needed to be fast and maneuverable, so Hawkins and his contemporaries had added new features to the traditional galleon, including split sails, faster methods of striking topsails, capstans for weighing anchor more quickly and so on.

English onboard-cannon technology was also advanced. There

were four kinds of guns. Cannons fired stone and heavy shot to damage an enemy ship, demicannons fired midsized shot, perriers were used for short-range midweight shot, and culverins fired light shot·long-range. The English had a total of 172 ships, carrying a total of 1,972 guns. They were lower in the water, with slimmer, longer hulls designed to move in and away quickly and to pound the Spanish with shot from a distance.

The 130 Spanish ships, carrying 1,124 guns, were old-fashioned carracks, normally used as escorts for the India spice galleons and to guard South American treasure fleets. Carracks were high-castled, lumbering tubs, built to swing in alongside an enemy ship so that soldiers could then board her. There were also galleys, unable to carry guns because of their oar ports. In general, because of their design and tactical approach to engagement, the Spanish ships carried heavy cannons, which worked best only at close range.

In the final tally, the English did more damage with their long-range culverins, staying well out of range of the Spanish cannons and denying the bigger Spanish ships their close-quarters advantage. The Spanish were also wrong-footed by English tactics, as Howard and Drake swept in a line around the wings of the Spanish fleet formation to rake them with fire from both sides. This maneuver finally split up the Spanish fleet and routed it. Stormy weather finished the job, driving the Spanish out of the English Channel into the North Sea and up around the northern tip of Scotland, to wreckage and death on some of the most inhospitable coastline in the world.

Philip's invasion had failed, but it had taught the princes of Europe the value of a navy. So by the mid–seventeenth century the English (and then everybody) were building gigantic, two-thousand-ton ships. The construction of one of these monsters used over a thousand oak trees, and soon there were laws aimed at preserving the dwindling forests for navy use. In Spain, for example, nobody was permitted to cut down a tree without planting two in its place.

Above all, stringent measures were adopted everywhere to restrict the use of trees for charcoal (which up to that point had been the principal industrial fuel). The new legislation was extremely bad news for charcoal-using glaziers because at the beginning of the seventeenth century, the market for glass was booming. The

European economy was moving into high gear, thanks to increasing imports from newly discovered lands to the east and west, the early beginnings of industrial development and advances in agriculture. Somehow a new fuel had to be found, and by 1612 there was already a technique (developed by an Englishman named Thomas Percivall) for making iron and glass with coal. This was done in a special, covered pot that protected the glass from the sulfurous fumes of the coal.

New kinds of glass soon appeared. In 1673 the London Glass Sellers Company engaged George Ravenscroft to develop flint glass, so called because it included silica, ground fine from flint. But the covered pot being used in coal furnaces made it harder to melt the glass mix thoroughly. Stirring the material meant removing the pot lid, which exposed the glass to impurities in the coal fumes. Some new ingredient had to be found to make the mix fuse more readily. It was probably Ravenscroft who hit on lead oxide. The result was noticeably clearer glass, free of the tiny cracks which had previously characterized the product.

In 1693 Louis de Nehou was set up in business by the French government at St. Gobain, near Laon, in northern France, to produce the new glass in large, cast sheets. His particular technique (first developed in 1675) involved pouring molten glass onto a stone table with raised metal edges (to prevent spillage), then flattening the glass with rollers and letting it harden for ten days. At that point the plate would be ground flat with another piece of glass and polished with a fine-ground stone powder. The new plate glass was just what Louis XIV[264] (who called himself the Sun King) wanted for his palace at Versailles, outside Paris. Here he ordered the addition of a great new reception room, running down one side of the building and lined all the way on both sides with full-length mirrors that Nehou's plate glass made possible. By day, everyone — the servants passing to and fro, carrying food or leading cows to provide the royal children with milk; the king's supplicants in their sedan chairs; and the foreign visitors in outlandish costume — saw themselves everywhere they looked. At night, the rooms were lit by the glow of hundreds of candles, reflected again and again. Louis's new "Hall of Mirrors" was the sensation of Europe.

The new, flat mirrors were also about to change life for maritime explorers by making navigation more precise. The problem was

264 10 18
264 274 235

that the explorers were bringing back cargoes that boosted the French economy and helped to pay for royal extravagances like Versailles. But they were sailing uncharted waters to do so, and their major obstacle was finding their way home with all the profits. At the time, navigation out of the sight of land involved a technique, developed earlier by the Portuguese, called "running the latitudes."[265] Navigators identified the ship's location (to the north or south) by checking the angle of the Pole Star, which rose higher in the sky as the ship sailed north.

Finding the position to the east or west involved a more complex calculation, in which the navigator measured the angle of two or more heavenly bodies and then checked in a star table what those star positions would be, at that hour, back in Greenwich, on the meridian line. From 1761 the new Harrison chronometer[266] indicated Greenwich time to within a fraction of a second. The difference between where the star would have been at that precise time in Greenwich and the time and angle at which it was now being observed on board ship told navigators how far around the world from Greenwich they were.

Apart from the need for accurate timekeeping, the more exact a star fix, the more precise the navigator's work would be. Star finding was made precise by the new, flat mirrors because they provided an undistorted image. In 1731 John Hadley, in England, and Thomas Godfrey, in Philadelphia, had independently developed a new navigational instrument, called a sextant,[267] that depended on a good mirror for its accurate operation. The instrument carried a telescope a few inches long, pointed at a small, half-silvered mirror. First, the navigator found the relevant star by moving a pivoted lever attached to a small mirror. At a certain point, the mirror caught the image of the star and reflected it in the silvered part of the glass fixed in front of the telescope. Through the unsilvered part, the navigator could see the horizon. When the swiveling arm had tilted the star-fix mirror so that the star's image was superimposed exactly on the horizon, the other end of the swiveling level pointed at a scale showing the angle of the star up in the sky.

But the sextant did one other clever trick. It could be turned on its side and used to make precise angular measurements. With this technique, two Protestant refugees, one from France (Joseph Desbarres) and the other from Holland (Samuel Hollandt), revolutionized the world of maritime cartography in the mid–eighteenth century. Desbarres had become a lieutenant in the British Royal American Regiment in 1756 and for a number of years was employed as a recruiting officer for the regiment in America. There he persuaded the British army to adopt the guerrilla tactics used by the Indians and backwoods French-Canadians whom the British were fighting at the time. During the Quebec campaign Desbarres was engineer for General Wolfe and carried out extensive surveys, as well as producing river charts and other maps in preparation for the Battle of Quebec. It was during this work that he met and began to work with Hollandt, an engineer who had been in the regiment since 1754.

Around 1764 the British high command decided, with unusual foresight, to chart the eastern coastline of Canada and the American Atlantic colonies. Desbarres and Hollandt were appointed to this task, and for ten years they sailed up and down the coast using their sextants with great enthusiasm. The work consisted of rowing ashore and measuring a straight line between two prominent features, such as headlands. Then they rowed a measured distance back out to sea, to a point where the sextant was used in horizontal mode to establish the angles between this offshore point and both coastal features. This triangle was the basis for another, similar triangle, which was then used as the base for another and so on, up and down the coast. Each time, the use of the sextant in its vertical, navigational mode also provided a very precise star fix, so that the exact location of each triangle could be established.

The two men finished their great work, which was finally published as *The Atlantic Neptune*, in the form of 15 large maps or charts engraved on 290 copperplates. The charts included soundings, as well as topographical descriptions like "rocky," "hilly make of country" and so on. There were many attractive views of the coast, painted as insets. When it was finished, the *Neptune* showed every detail of the American eastern-seaboard coastline, south all the way from Nova Scotia to Florida, round to the Mississippi Delta. The charts would have provided vital information to the quartermasters supplying the British troops at war with the American revolutionaries. However, the *Neptune* was published only weeks before the War of Independence ended in American victory.

All was not doom and gloom. In Canada one of the more assiduous pupils of Desbarres and Hollandt had been a young naval officer called James Cook,[268] who had helped with the survey of the area   268 163 *128* around Quebec and who used the knowledge he had gained from the two cartographers in order to chart (and plant the British flag on) the coasts of Australia and New Zealand. Win some, lose some.

The only thing now missing from these new accurate charts and maps was information on the height of the terrain. Inexplicably (given that the barometer had been in general use for at least fifty years), it was not until 1787 that a Swiss professor of philosophy at the University of Geneva would do anything about it. Henri de Saussure had an obsession with Mont Blanc. He even offered a prize for the first person to climb it (although when the peak was scaled in 1786, he refused to pay up). The following year he became the third person to climb to the summit, accompanied by eighteen helpers carrying his many scientific instruments. These included a barometer[269] with which he was able to confirm that the mountain was 4,807 meters high. This information immediately set the style for mapmakers, and all maps carried height information from then on.

269 171 *135*

Although Saussure spent only four and a half hours at the top of the mountain, he was able to conduct dozens of observations and a series of experiments that excited the public's imagination (and for which he was elected an honorary Fellow of the English Royal Society). The Swiss were so proud of him there were even proposals that the mountain be renamed Mont Saussure in his honor. Later on, in 1796, he would write up his experiences in a four-volume work called *Journeys in the Alps*, which virtually single-handedly created Swiss alpine tourism and a worldwide interest in downhill skiing.

Many further alpine expeditions gave Saussure the opportunity to develop ideas on geology, particularly on the formation of the mountains all round him in Switzerland. At the time there were two main schools of thought regarding mountains. One view, called Neptunism, held that they were elevated bits of ancient seabed and that eons in the past the Earth had been covered by a universal ocean. The other, the Plutonist view, thought the interior of the Earth to be of molten magma, which periodically burst out in volcanic eruptions that made mountains. These were then eroded over time and worn down, forming sedimentary layers in the sea. Both views accounted for the fossils that climbers like Saussure were beginning to find on mountaintops, although the Plutonists' erosion process presupposed an Earth much older than the five- or six-thousand-year-old planet described in the Bible.

At the end of his *Voyages* Saussure added his "Agenda," listing what he described as the key questions still to be answered regarding geological processes and the age of the Earth. It may have been Saussure's example that stimulated an unusually observant Scottish farmer and amateur geologist, James Hutton, to develop what became known as uniformitarianism. This is, in essence, a theory that presupposes geological processes to have taken the same length of time in all epochs, so that present-day geologic events faithfully reflect the sequence of similar events in the past. In 1785 Hutton wrote a paper, which was delivered at the Edinburgh Royal Society by his friend Joseph Black,[270] the man who had shown James Watt   270 219 *184* how to improve the steam engine. The paper was entitled "Ascertaining the age of the world, its past changes and likely future." In it, Hutton concluded: "It has required an indefinite space of time to have produced the land as it now appears. . . . An equal space has been employed upon the construction of that former land from whence the minerals of the present came. . . . There are presently lying at the bottom of the ocean the foundations of a future land which is to appear after an indefinite space of time."

As a writer, Hutton made a good farmer; so it took the publication of a concise version of his theories by John Playfair in 1802 (called *Illustration of the Huttonian Theory of the Earth*) for Hutton's and Saussure's ideas to reach a wider audience. One of Playfair's most avid readers was the geologist Sir Charles Lyell, who in turn synthesized the uniformitarian view in his own work. Finally, in this chain of rewrites, it was Lyell's book *Principles of Geology* which moved Charles Darwin to say that when he read it "the scales fell from my eyes." Darwin's realization that the Earth might be geologically old enough for immensely slow, natural processes to occur confirmed his theory of evolution, and in 1859 he published it as *Origin of Species.*

Key to the next step (in this story of individuals and the way they affect what happens on the web) was Darwin's view that when environmental change occurs, those species that can adapt to fit the new conditions best, survive best. This adaptability, he thought, is a hereditary process that transmits to later generations the strengths that will help them survive or the weaknesses that will lead to their extinction. Darwin expressed his belief that inbreeding is especially inimical to the success of this process, saying, "When relations unite . . . there is a decrease in general vigour."

Ten years after Darwin published his groundbreaking work, his cousin, Francis Galton, took up the matter of hereditary abilities, with an exhaustive analysis of eminent people's obituaries in the London *Times*. He claimed that his study showed there were only 250 "eminent" people per million. For Galton, this gave the lie to "pretensions of natural equality." Galton also calculated that in the population at large there was roughly the same percentage of imbeciles as eminent persons. Concentrating on judges born between 1660 and 1685, Galton concluded that those who had married heiresses were less fertile because heiresses (female children) themselves were a sign of infertility in their families, and this trait was passed on. Judges also obviously passed on their qualities of "eminence," because even those of their children who did not become judges attained eminence in some field.

Galton went on to study the lives of 180 eminent men of science (they were all Fellows of the Royal Society) and compared their data with those for the same number of ordinary male citizens of the same age-group. He found the ratio of eminence to ordinariness to be 1:10,000 and decided to find out why only one man out of ten thousand should stand out in this way. Galton decided the evidence proved that the large majority of eminent persons were "racially pure" (both parents English, or both Scottish, for example). The largest number of eminent British people came from southeast Britain, and the smallest from the northeast and the Scottish Highlands. Most eminent people were also the eldest sons of large families and were generally more energetic, healthier and taller than average.

In 1901 Galton brought his work to its logical conclusion with a lecture to the London Anthropological Society entitled "On the possible improvement of the human breed under the existing conditions of law and sentiment." In it, he made public a word he had coined to describe his theory: eugenics. Three years later there was a Eugenics Society where lectures were given with titles such as "Restrictions in marriage."

From 1907 Karl Pearson occupied the Galton-endowed Professorship of Eugenics at the University of London and set the stage for the implementation of Galton's ideas in ways that neither Galton nor the evolutionary cousin who had inspired him would ever have dreamed possible or desirable. Pearson strongly believed that the higher birthrate of the poorer classes was a threat to civilization, and he expressed the kind of racialist views which were to be taken up with vigor in the U.S. immigration quota laws of 1924. At the time, the country was gripped by fear of criminal and defective immigrants flooding the country. The new quotas limited immigrant numbers to 150,000 a year with no more than 2 percent representing each nationality. By 1935 there were laws in twenty-seven U.S. states, as well as in Denmark, Switzerland, Norway and Sweden, permitting the sterilization of idiots, epileptics, the insane and, in some places, the merely criminal.

The eugenics movement reached its apotheosis in Nazi Germany. In *Mein Kampf*, Hitler was to write: "It is the Aryan alone who founded a superior type of humanity; therefore, he represents the archetype of what we understand by the term 'Man.' . . . The state must assert itself as the trustee of a millennial future . . . to avail itself of modern medical discoveries. It must proclaim as unfit for procreation all those who are afflicted with some visible, hereditary disease, or are the carriers of it; and practical measures must be adopted to have such people rendered sterile." In 1934 compulsory sterilization was legalized in Germany.

But what Francis Galton had really been looking for all along, in his research into the origins of differences between people, was the secret of why one individual was uniquely different from another and how this difference could be identified. The last thing Galton would have wanted was to see his theories perverted in the wholesale genocide practiced by the Nazis.

Galton ultimately discovered what he was looking for because he found a way to describe each person's uniqueness. He did so with a technique that brings this chapter back to the billiard ball with which it begins: In 1890 Galton took up a matter he had heard about a few years before from a doctor named Faulds, a medical missionary in Japan who had written to Darwin to say he thought there might be racial differences identifiable in fingerprints.[271] Darwin had passed the letter on to Galton, who had passed it on to the Anthropological Society, where it lay forgotten because Faulds's social standing warranted little interest in what he had to say. But

271 190 *154*

when Sir William Herschel, ex–commissioner of Bengal, wrote to Galton about his use of prints to prevent impersonation and fraud, Galton searched out Faulds's letter.

After collecting more than 2,500 ten-finger sets of prints, Galton realized that the distribution of patterns was peculiar to each finger and that there were three distinct shapes to the patterns: whorls, loops and arches. The possible number of combinations of patterns on all ten fingers of one individual gave an overall set of patterns that was virtually unique. At the end of the nineteenth century, Edward Henry, chief of police in Bengal, visited Galton's lab and produced a variant on the system which would make it foolproof. In essence, the Galton-Henry technique consisted of drawing a reference line between two characteristic points on a fingerprint and counting the number of pattern lines it crossed. In this way, the chances of two individuals having the same reference-line crossing number was one in ten trillion.

In 1902 the system was taken up by Scotland Yard, in London, where it was first used to catch a petty thief, Harry Jackson, who was identified by the fingerprints that detectives found on what he had stolen. His haul had been a set of billiard balls.

Fingerprinting has been a forensic technique of prime importance in the fight against crime all over the world. Therefore, it is ironic that the first international law came into existence because of a criminal act. . . .

17 Sign Here

O NE of the things you'll have noticed, if you're one of the ever-increasing number of first-time international travelers, is that you need to watch out for differences in local law. It's very much a case of "different strokes . . ." In some places, for instance, you can't import magazines with pictures of naked people in them. In others, you can be executed for bringing in drugs. Elsewhere, you might get your hand chopped off for theft. What's pornographic in one country may not be so in another. In certain U.S. states (but in no European country) it is legal to turn right on a red light. And as information technology brings everybody closer to one another, the complexities of local law become more and more frustrating, especially in regard to, say, intellectual property rights.

One of the few places where this confusion does not occur is on the open sea. There all are free to come and go as they please. Curiously, this admirable example of international legal cooperation began in 1604 with a crime. In the Strait of Malacca, off the coast of Malaysia, a warship working for the Dutch East India Company captured a Portuguese merchant ship, the *Santa Caterina*, bound for home with an extremely valuable cargo of spices. When the captured galleon was brought back into Amsterdam harbor and her cargo sold for considerable profit, the event caused an international outcry. There were even voices among the company shareholders arguing that this attack on fellow Christians was reprehensible and should have been disavowed by the Dutch government.

This reaction may have been the reason why the company hired a sharp young lawyer, Huig de Groot (Hugo Grotius), to express an opinion on the case. His response took the form of a short treatise on the subject of spoils at sea. In a nutshell, de Groot said that the sea was common to all, since it was not actually "property." Because it was so limitless, it could not properly belong to anybody. Also, it had never been occupied, and all rights of possession arose from occupation. Nor could the Portuguese lay claim to the Indian Ocean as they did, since they had not been the first to sail on it (de Groot cited the cases of Indians and Alexander the Great). Further, the fact that the pope had "given" the Portuguese the harbors of the East Indies was "mere ostentation," since they had not been the pope's to give. The ocean was free for all men to trade on, held de Groot, and if the Portuguese continued to use arms to back their claim to a monopoly of routes to the Eastern spice islands, the Dutch were justified in opposing them.

In 1609 de Groot published all this as *Mare Liberum (The Open Sea)*; although individual monarchs might have quibbled about the details, in general its approach was welcomed. The exploration of Africa, America and the East had created a confusing hodgepodge of quasi-legal claims and counterclaims. Europeans were planting their flags on virtually any piece of coastline dry enough to land on, the pope had given Brazil to the Portuguese even before it had been discovered, Protestant explorers refused to recognize Catholic laws overseas and the Turks laid claim (modestly) to the entire planet. About the only people not involved in staking claims were the native peoples whose lands and resources were being grabbed by colonialists. So when, in 1625, de Groot published his masterpiece, a work called *On the Law of War and Peace* (and the first systematic treatment of international jurisprudence), he established international law almost overnight. De Groot's work immediately conferred order and process on the business of trade and gave a tremendous boost to the financial fortunes of the newly emerging nation-states of Europe.

One of the happier effects of this economic boom was that there were soon large amounts of money available, as commercial banking grew rapidly and, with it, public finance. The complexities of this new high-finance world made life difficult for those who now had to calculate large numbers. So it was in order to help his father (a tax collector for the French government) that in 1642 a young

mathematician named Blaise Pascal[272] designed a counting machine. 272 169 *134*
It worked much like an old-fashioned taxi meter did, with toothed
wheels and gears, moving drums carrying numbers that appeared
in small windows cut into the casing.

Pascal's machine could add and subtract, as well as multiply and
divide (by repeated addition or subtraction). In the end, he built
fifty of these machines, made of wood, ivory, leather, ebony or metal.
Everybody (including Descartes and Queen Christina of Sweden)
was most impressed, and in 1649 Pascal was granted exclusive
manufacturing rights in France. Unfortunately, the calculator was
too expensive, and the venture fizzled. Nonetheless, this was a time
when it was hard for a good mathematician to put a foot wrong.
Numbers were all the rage. Nations were waking up to the value of
measuring the size of their populations so as to plan taxation and
expenditure better. The risks of exploration in distant, uncharted
seas triggered a healthy demand for maritime insurance; the idea
spread to include fire, and then life. Life insurance got a great boost
when the Englishman John Graunt[273] published actuarial studies of 273 159 *126*
birth-and-death parish registers, showing how to calculate a person's probable length of life at any time. The data made it much
easier for insurers to calculate premiums more accurately.

Probability was to be the key to Pascal's brief life (he died at
thirty-nine), thanks to the fact that the other contemporary craze,
besides statistics, was gambling. By the mid–seventeenth century it
had become the principal leisure-time activity of impecunious
French aristocrats, who were forbidden by law from making any
money in commerce. The boom generated by international trade
was passing them by, and gambling was an honorable (if illegal)
alternative way to make a fortune. From 1652 to 1654 Pascal spent
time looking into the mathematical aspects of betting, probably
because he was trying to work out a successful system for an aristocratic gambler friend. Among other things (in case you're that way
inclined), Pascal worked out that to be sure of getting a pair of sixes
from two dice would require 24.555 throws. Legend has it that his
friend went on to make a fortune.

Pascal then addressed the other problem facing gamblers whose
game might be broken up at any moment by the authorities. The
problem was how to divide the wagers laid so far by the various
players and based on the probable outcome of the game. Pascal's

answer was to find a way of showing the most probable outcome of any event with his "arithmetic triangle." It looked like this:

$$
\begin{array}{ccccccc}
& & & 1 & & & \\
& & 1 & & 1 & & \\
& & 1 & 2 & 1 & & \\
& 1 & 3 & & 3 & 1 & \\
1 & 4 & & 6 & & 4 & 1 \\
1 & 5 & 10 & & 10 & 5 & 1
\end{array}
$$

Say the triangle were used to calculate the chances of the next round of coin-tossing, in a game where three coins were being used. The third row of numbers (the top number not constituting a row) contain 1, 3, 3 and 1. These add up to eight, so this third row indicates the number of ways three coins could fall (i.e., eight ways): once as three tails, three times as two-tails-and-a-head, three times as two-heads-and-a-tail and once as three heads. The fourth row, which provides the same calculus for four coins, adds up to sixteen. This is the number of ways four coins could fall. And since the triangle is infinitely expandable, the probabilities could be calculated for an infinitely large number of coins. With his triangle, Pascal laid the foundations for the calculus of probability.

But it was to be probability of a very different kind that would influence his life much more drastically than his time at the gambling tables. It all came down to the company his sister Jacqueline kept. In 1654 she entered a rather unusual nunnery in Paris, called Port Royal, the center of French liberal thought at the time. The following year Blaise joined her for three weeks of meditation, prayer and conversation and was immediately attracted by the kind of talk he heard there. The nunnery was an intellectual power-house, serving as a retreat for thinkers from all walks of life, principally lawyers and scientists, all of whom held Jansenist views.

Cornelis Jansen had been a theologian at the University of Louvain, in the Catholic southern part of Holland, and had led a movement for church reform which ended up arguing with the Jesuits about matters of probability. Jansen's great work was a treatise on Saint Augustine's thought, in which he argued that the contemporary church needed spiritual reform. There was too much emphasis on reason, said Jansen, whereas experience alone should be the individual's spiritual guide. No amount of churchgoing would save a soul, only the love of God. Conversion to the faith happened only if God wished it; and no schooling or catechism was needed when conversion happened, because its effect was instantaneous.

Others soon took up this cause, and by 1653 the pope was declaring Jansenist propositions to be heretical because they questioned the church's authority over individual behavior. The Jesuits held that the church was infallible, so that if an individual were unsure about whether an action was sinful or permissible, the church could provide arguments as to which was the more probable case. In matters of conscience, therefore, a believer might safely follow the opinion of a recognized doctor of the church. The Jansenists held the opposite view: that in matters of conscience, it was more probable that if there were any doubt about an action, the safe thing to do would be nothing — in spite of anything the church might say on the issue. For Jansenists, it was more probable that this would be the right course of action. For these reasons, the issue boiled down to a fight between Jesuit "probabilism" (holding that the church was probably right) and Jansenist "probabiliorism" (arguing that individual conscience was *more* probably right).

In 1634 a noted Jansenist took over as spiritual director of the Port Royal nunnery, and by 1643 the little community of liberal thinkers had begun to grow. In time, the group set up a school and began to publish textbooks, one of which was a highly critical piece on Holy Communion that sold out in two weeks. To add fuel to the flames, between 1656 and 1657 Pascal published his anonymous satirical *Lettres Provenciales*, attacking the Jesuits and their probabilism. The church reacted with the full threat of papal interdict. At the insistence of Rome, Louis XIV[274] closed down Port Royal, and by 1666 most of the community was dispersed across France or in hiding. In 1709 the building itself was destroyed. Even the nuns' cemetery was dug up, and their bones exhumed and scattered.

274 10 *18*
274 264 *223*

But the Jansenist movement continued to attract adherents all over Europe. About thirty years later, in Paris, a young priest, Charles-Michel de l'Epée, refused to sign a declaration opposing Jansenist doctrines and was deprived of his ecclesiastical position. Casting around for work, he found a couple of young, deaf sisters previously in the care of a priest (who had recently died). L'Epée took on the responsibility for their religious instruction and, in order to communicate with them, worked out a set of simple hand signals. This was not a new idea, since sign language had been in use in religious communities as early as A.D. 909.

The practice had originated with monastic orders like the Trappist Benedictines, who enforced silence on their members. This had some amusing and unforeseen effects, because sign language was all too often far from contemplative. In 1180 the chronicler Gerald of Wales described the atmosphere one evening when he dined at the high table of the Canterbury Cathedral priory: "There were the monks ... all of them gesticulating with fingers, hands and arms, and whistling to one another instead of speaking, all extravagating in a manner more free and frivolous than was seemly; so that I seemed to be seated at a stage play or among actors and jesters." By the thirteenth century, sign-language vocabulary in some monasteries was extremely sophisticated, even making possible complicated theological discussions in silence. One German abbot, Wilhelm von Hirsau, listed 359 separate signs in use at his abbey.

In 1607 Juan Bonet, a Spanish soldier of fortune, entered the service of Juan Fernandez Velasco, constable of Castile, whose son was deaf and whose deaf uncle had been taught sign language by monks. In 1620 Bonet published the first book on teaching the deaf and called it *The Reduction of Letters and the Art of Teaching the Mute to Speak*. The book was a manual for one-handed signing, and sometime in the 1750s l'Epée obtained a copy, set up a school in Paris and went on to develop his own system. His system was so easy, he claimed, that a student could learn the signs for more than eighty words in less than three days. Training began with the basic manual signs for the alphabet, and then students were shown pictures of objects and the appropriate signs. L'Epée's school earned such a reputation that at one point the papal nuncio in Paris watched as a student answered two hundred questions in three languages. In 1786, when the bishop of Bordeaux wanted to establish a school for the deaf, he chose a priest called Sicard, who was trained in l'Epée's methods, to be the new school principal. Sicard was so successful that three years later, on l'Epée's death, he took over in Paris and completed l'Epée's dictionary of signs. In 1791 the Paris school became the National Institute for Deaf Mutes, survived the ravages of the French Revolution and went on to achieve an international reputation under Sicard, who was by now a chevalier of the Legion d'Honneur.

In 1815 Sicard received an American visitor who wanted to be trained in the l'Epée system: Thomas Gallaudet. Earlier, on vacation from Andover Theological Seminary in New England, Gallaudet had met and taught the deaf daughter of a prominent local surgeon. The man was so pleased that he raised the money needed to send Gallaudet to Paris so that he could learn l'Epée's techniques at Sicard's school. By 1817 Gallaudet was back in Hartford, Connecticut, for the opening of the first American free school for the deaf.

Gallaudet eventually married a deaf woman and they had a son, Edward. After several failed attempts at being a missionary, and not being able to afford to go to China, Edward settled for teaching the deaf. He started a school in Hartford and then, at the age of twenty, was appointed to run the new Columbia Institution for the Deaf and Dumb in Washington, D.C. The school had been endowed with a house and two acres of land by Amos Kendall, a lawyer who became the U.S. postmaster general. Kendall subsequently served as Samuel Morse's business manager, after acting as an intermediary between him and Congress when Morse[275] was trying to get government backing for his telegraph.

275 30 29
275 114 85
275 235 197

Kendall persuaded Morse to develop the telegraph with private funding, and the two men agreed to split the profits, with Kendall receiving 10 percent of the first $100,000 and 50 percent thereafter. When Morse set up his first telegraph line between Baltimore and Washington, it ran across Kendall's land. Given the extraordinary commercial success of the telegraph, Kendall became very rich and decided to donate land for the Columbia Institution because Morse's wife was deaf and Morse had thought of the Morse code, he said, by tapping messages on her hand. The institute was to become the only educational establishment in the world to offer degree courses to the deaf and dumb.

A few months after Gallaudet took over the institute, a controversy erupted in America about whether deaf people should be taught to use sign language or rather taught to speak (an approach known as oralism). This argument about what to teach the deaf would eventually lead to the American development of the telephone. The sequence of events that led to it began when an immigrant Scottish elocution teacher called Bell, whose wife was deaf and who, in the course of teaching actors to improve their diction, had invented a series of simple drawings to show the position of the speech organs during articulation. In 1867 he described the system in a book entitled *Visible Speech: The Science of Universal Alphabetics.*

In demonstrations of how well the system worked, he would often ask his sons to enter a room where they would be shown deaf-and-dumb signs for words (in English, French or Gaelic) or for noises like hissing or kissing. The boys would then reproduce the words and sounds with extraordinary precision. So when the principal of a London school for deaf children asked if she might use visible speech, one of the sons, Alexander (now grown-up), taught at the school with considerable success. After only one lesson the children were uttering intelligible sounds. By the time Alexander emigrated first to Canada and then to the United States, he was a firm supporter of oralism, even during a two-month teaching session at the signing school in Hartford where Gallaudet had started.

By 1872 he was teaching vocal physiology at Boston University and teaching deaf students in his spare time. Searching for a visual method of improving his deaf pupils' voice control, he came across an instrument called a phonautograph, invented by a man called Leon Scott. When a sound was made into a horn, the noise vibrated a membrane fitted across the other end of the horn. Attached to the membrane was a stiff bristle, the free end of which moved up and down in response to the vibration and traced out a pattern on smoked glass.

It was while working with this technique, in 1875, that Alexander had the idea of fixing a small magnet to the membrane, so that the vibration would move the magnet in and out of a coil of copper wire sitting next to it. This movement would generate a varying current of electricity in the coil, and the current could then be sent down a wire. At the other end, the varying signal would enter another coil of copper wire set next to a small magnet on a membrane in a second horn. The varying current would cause the magnet to move back and forth, vibrating the membrane to reproduce the original sounds. Thanks to this idea, Alexander Graham Bell[276] became the inventor of the telephone.

276 34 *30*
276 54 *238*

Leon Scott's interest in representing sounds with wiggly, phonautograph lines was part of a general trend in science to find better ways to present information in graphical form, so as to make medical data clearer and easier to analyze. Early-nineteenth-century Paris hospitals had benefited from the fact that they were packed with thousands of wounded from the Napoleonic Wars and thus could develop statistical techniques for analyzing symptoms and assessing treatment. Throughout the rest of the century, more attempts would be made to turn data like temperature and blood pressure into graphs from which physicians and nurses could update reports on a patient's condition.

By mid-century medical technology had produced instruments for gathering data, including thermometers for temperature, stethoscopes for lung and heart noises, the hemodynameter[277] for the way blood pressure goes up and down with breathing, the mercurial pump for collecting air from blood, as well as ophthalmoscopes, otoscopes, laryngoscopes, endoscopes and sphygmographs (for recording the pulse). Early advances had been made by Carl Ludwig, professor of physiology at Marburg, Zurich, Vienna and Leipzig. In 1847 he had introduced the graphical method with an instrument called a kymograph, based on a float resting on mercury. As the mercury rose and fell with changes in respiration, the float rose and fell, moving a stylus that recorded the changing mercury levels on a strip of paper rotating on a drum.

Etienne Marey, a French doctor, extended these techniques to include graphs of blood pressure, heartbeat, respiration and muscle contraction in general. Marey's most famous invention (still in use until 1955) was his tambour: a small, air-filled metal capsule, covered at its open end by a rubber membrane. When any kind of compression (such as that caused by movement) depressed the membrane, air was forced out of the capsule and along a rubber tube to another, similar capsule covered by another membrane and carrying a stylus which then traced the alterations in pressure.

One of the less specifically medical uses of the tambour was associated with the upsurge of interest in languages that had begun at the end of the eighteenth century with the work of people like William Jones and Franz Bopp. Both men published Sanskrit[278] grammars, thrilling Europeans with talk of an ancient ancestral Indo-European tongue. This discovery got people interested in the ways in which all European languages descended from their ancient progenitor, which in turn triggered investigations into the physiology of language. As part of these investigations, tambours and

277   183   *145*

278   312   *283*

similar instruments were placed in people's mouths to measure what was happening to the various speech organs when they spoke. These early experiments were aimed at developing the scientific analysis of speech production.

Near the end of the nineteenth century, this research began to have wider implications as the first international organizations were formed. In an era before simultaneous translators in booths, people became aware of the need for better methods of communication between delegates. One suggestion, by the Swiss Ludwig Zamenhof, was to go back to a common language again. So in 1887 he invented one called Esperanto, based on word roots shared by all Indo-European languages (at the time spoken by twice the number of those speaking the next-largest language group, Sino-Tibetan). In spite of its vaguely utopian aims, Esperanto never caught on, possibly because other ways were found to aid international understanding. One of these was the development by linguists of a new kind of alphabet that made writing any language accurately a relatively simple matter.

The International Phonetic Alphabet first saw the light of day when it was published in 1897, but it owed its existence in great part to the mid-century phonography experiments conducted by Isaac Pitman. Like Zamenhof's, Pitman's interest in phonetics sprang from his internationalist utopian leanings. He began, however, with English. Part of his great plan was to make his fellow Englishmen write their language phonetically and, in general, to develop simpler symbols for writing than alphabetic letters. In 1850 he set up the Phonetic Institute, which gave birth to the British Phonetic Council. (Alexander Graham Bell's father was a member.) Pitman's efforts failed but did lead to the shorthand system that today carries his name and that did eventually help to inspire the fully fledged International Phonetic Alphabet. One of the early pioneers in phonetics was Henry Sweet, who was the model for Shaw's character Henry Higgins, in *Pygmalion* (later musicalized as *My Fair Lady*). In an ironic twist, Higgins uses Bell's visible speech symbols in the play to record Eliza Doolittle's speech, and she notices the fact ("That ain't proper writing!").

But now that the phonetic alphabet made it easy to reproduce and analyze any language — no matter how complex — it opened up whole new areas of study. These interests were associated with the "new" science of anthropology, the Romantic continent-wide

craze for ancient folk origins, the new sense of nationalism stimulated by the breakup of the Austro-Hungarian Empire and so on.
People had suddenly started to become intensely aware of how
different they all were. So, armed with the phonetic alphabet, linguists headed into dialect country.

Dialectologists (most of them German) sent teams of researchers
bicycling around southern Italy to take notes on the local speech
differences or handed out fifty thousand questionnaires asking Germans to pronounce the sentence, "In winter the dry leaves fly
around in the air." One typical aim was to gather enough data to
draw maps showing the boundaries between places where Germans
said *dorf* rather than *dorp* ("village") or where the word for "I" was
variously pronounced *ich* and *ick*. One of the linguists involved in
these capers was Jena University philologist Edouard Schwan, who
became *the* German expert on the French accent. At one stage in
his research, he solicited the help of one of Germany's more eminent
physicists, Ernst Pringsheim, whose contribution was to carry out
a phonometric analysis of the accent, using what was probably a
variant of Scott's phonautograph.

However, Pringsheim[279] was considerably better known for his
work on radiation, in which he first used the radiometer (the new
hi-tech wonder-instrument of physics) for the measurement of infrared light. Unfortunately for Pringsheim, his radiometric reputation didn't last long — it eventually turned out that the radiometer
wasn't doing what he thought.

279 88 *60*

The instrument had been invented by one of Victorian England's
most respectable scientists, William Crookes, who was motivated
to seek ways of using science to make money because he had to
support a wife and ten children. One of his ideas, a new process for
gold extraction, made him financially independent, and he was then
able to make some genuinely valuable contributions to science. He
invented the cathode ray tube,[280] as well as the first dry photographic process, and helped determine the atomic weight of thallium. But his invention of the radiometer perhaps owed less to his
scientific abilities than to his somewhat eccentric interest in the oc-
cult. Crookes was a regular participant in seances at which he
claimed to have seen, among other things, suspended handkerchiefs,
self-playing accordions and, most frequently of all, a ghost named
Katie King. And all this while he was president of the Royal Society.

280 40 *32*
280 50 *41*

The idea of the radiometer first occurred to him during his

research into thallium. He was using a balance in a vacuum, and he noticed that warm samples appeared to be lighter than cold ones. At first he thought there might be a link between heat and gravity, and he continued to investigate the phenomenon because he was also interested in the possibility of "psychic forces." He found that if a large mass were brought close to a lighter one, while both were suspended in a vacuum, the two masses either attracted or repelled each other. And the stronger the vacuum, the greater the effect. By 1873 Crookes was convinced that he had discovered the repulsive effect of radiation.

So he designed the radiometer to investigate this effect. The device consisted of four arms suspended on a steel point resting in a cup. To each arm he fixed a thin vane of pith, with the same side of all four vanes blackened. When a light was held close to the vessel, the little vanes moved away from the light, causing the radiometer arms to spin. Crookes announced that the radiometer was reacting to the "pressure of light," and it was this assertion that would lead Pringsheim to his experiments with infrared radiation.

As it turned out, neither Pringsheim nor Crookes was right. In 1875 Osborne Reynolds, an English professor of math in Manchester (with the disconcerting habit of breaking off in the middle of lectures to write a new idea on the blackboard, much to the annoyance of his students), became the world's expert on flow. He was able to show that however close to absolute the vacuum in the radiometer might be, the heat from a light is actually causing the pith vanes to release a few molecules of gas, which then heat up and move, pushing against the vanes and causing the radiometer arms to spin.

Reynolds's interest in flow led him in a number of innovative directions: to develop ways of modeling river estuaries and canals, to produce patent improvements in pumps and turbines, to develop the math needed to make accurate scale models of ships, to discover cavitation caused by propellers, to study the calming effect of rainfall on the sea, to explain how a Ping-Pong ball stays supported by a vertical jet of water and to show how skis slide on snow because they melt it. Reynolds also came up with a zany theory that the universe is made of small balls.

But his greatest contribution to the sum of human knowledge is the number which bears his name today. The "Reynolds number" is a mathematical figure that says vital things about flow. The number is the product of a flowing material's velocity, density and diameter, divided by its viscosity. And the magic thing about the Reynolds number is that it tells engineers how to avoid turbulence. This is why ships move along smoothly, canals don't break their banks, pumps work efficiently and high-pressure gas and water come through pipes to your home without causing leaks. The Reynolds number has one other important use, without which much modern tourism and commerce couldn't do what they do. It tells engineers how to design airplanes because it is the basis of all aerodynamics. This was why the Wright brothers built a wind tunnel, studied airflow Reynolds numbers and tested three gliders before attempting powered flight.

But the fact that their airplane took off at all owed one other important debt to Reynolds. He had also done fundamental research on the effects of lubrication on one vital component that would be used in the Wright brothers' airplane engine. It was a device that, back in 1879, had originally been tried in bicycles. And the Wright brothers sold and repaired bicycles,[281] so they knew about it. The device that made both their bikes and their plane engine work was the ball bearing.

281 68 46
281 77 50

The man who told the world all there was to know about ball bearings was a German named Stribeck. At the end of the nineteenth century, he conducted the most exhaustive series of tests possible on every conceivable kind of bearing: cylindrical, spherical, those made of hardened and unhardened metal, in cups, in rings, on collars, on plates, in sets, alone, in grooves, under loads from all angles, spinning fast and slow, oiled and dry, hot and cold, smooth and scratched, large and small, or stressed by weights to the point of destruction. Stribeck found out that ball bearings work best if they are mounted in a grooved track, because this increases the load-carrying capacity of the ball.

It was ball bearings that kept Allied planes flying all through World War II. But Stribeck's work helped the war effort in one other vital way. A major difficulty encountered by bomber crews was that at high altitude, because of the lower air pressure, their fountain pens would sometimes squirt out excess ink all over navigators' maps. This was a serious inconvenience because it often meant that if essential points on the map were covered by ink blots, aircraft could get lost. So when a Hungarian living in Argentina offered the British and American air forces a way to avoid the problem, they leaped at it. His name was Lazlo Josef Biro, and in 1944 he arrived in England with his new invention, thirty thousand of which were bought by the British RAF, and even more than that by the United States Air Force.

By this stage of the war, thanks to the ball-bearing work of Stribeck and the aerodynamic work of Reynolds, Allied warplanes were regularly taking off on bombing missions. Now, also thanks to Stribeck and Reynolds, they could be confident of navigating to their targets and back again with maps clean and unblotted by spilled ink. Because Lazlo Biro had used Stribeck's data to design a device consisting of a pencil-shaped reservoir of ink feeding a small steel ball mounted on the end in a set of grooves. Biro had also used Reynolds's data on lubrication to control the way the ink flowed onto the ball and then onto a page. Thanks to both men, Biro's new ballpoint pen would never make blots.

Today, the ballpoint pen is so cheap, it's a perfect example of the throwaway item on which modern industries have been based for a hundred years. . . .

18 Bright Ideas

ONE of the really exciting things about the way change happens is how often somebody has a bright idea about something quite insignificant, triggering a series of events that ends with something of cosmic importance.

In 1892 William Painter, of Baltimore, Maryland, invented a piece of technology that was specifically designed to be insignificant. (He had already had a number of bright ideas and come up with a new railroad-car seat, a counterfeit-coin detector, a seed sower and assorted pumping devices.) Painter's little idea was one of the earliest examples of the kind of throwaway technology that keeps the economy of the modern world ticking.

The problem Painter solved with his gizmo had been around for some time: how to keep the fizz in fizzy drinks. At the time, this concern was no mere matter of keeping lemonade drinkers happy. The nation's health was at stake. This was all due to the fact that ever since Joseph Priestley[282] had invented soda water in 1777, effervescence was considered to be good for every infirmity. In an age of surging population, industrial city slums filled to bursting, rampant cholera and tuberculosis, no proper sanitation and doctors who thought all disease was generated by bad smells, carbonated liquids were as good a cure as any. Besides, real scientists claimed bubbles were the antidote to scurvy, so the British navy went for soda water in a big way.

The critical problem was to retain the sparkle, and thus the cure. Many methods had been tried, including cork and wire, cork and

282 5 14
282 161 128
282 306 278

string, cork and sealing wax, cork and tar, and glass marbles. None of them worked well. Then Painter invented his patented Crown Cork Seal, better known today as the bottle cap. The device was ideal for use with a brand-new drink that was enjoying runaway success as a "brain tonic." It was called Coca-Cola.

It was Painter's advice to one of his employees that triggered another great example of throwaway technology in the modern world. The fellow in question had the usual nineteenth-century young American businessman's start in life: clerk in Chicago, drummer in Kansas and scouring-powder salesman in Britain. Early in 1892 he joined Painter's salesforce, with responsibility for the New England states, but his own horizons were wider than that. In 1894 he gave a hint of the scale of his ambitions when he published a monumental flop called *The Human Drift*. The book outlined his plan to incorporate the entire world, with the express purpose of his setting up the Twentieth Century Company. This idea also flopped. Then, he said, he remembered something his boss at Crown Cork had once told him, "The secret of success is to make something that people will buy and throw away." As this philosophy had permitted Painter to retire rich after only ten years in business, it seemed like a good idea.

He cast around for an everyday activity that was so commonplace, it would create a permanent market for any relevant invention. One morning he was looking in the mirror when the bright idea hit him. As he put it, "If the time, money and brain power which we have wasted in the barber shops of America were applied in direct effort, the Panama canal could be dug in four hours." So King Camp Gillette invented the razor blade and made the modern world clean-shaven.

This feat was less simple than it sounds. Cutthroat razors had been in general use for nearly a hundred years. They were valuable objects, often considered family heirlooms to be passed on from father to son. For everybody but the poorer classes, however, shaving was not something one did oneself. Most people who shaved at all went to barbers. These craftsmen were Gillette's major opponents in his attempt to persuade the public to shave themselves, until they realized that barbershops would be the primary point of sale for the blades.

Gillette's success came principally from relentless advertising.

He said that he wanted his face to become as famous as George Washington's, so the company advertisement carrying his portrait was designed to look like a dollar bill. Gillette also refused to believe metallurgy experts at MIT who said that he couldn't put a sharp edge on a ribbon of steel. So he went into business with a man named Nickerson, who found how to harden the steel to make the edge possible. Within six months they were broke and in debt. But Gillette persisted, and in 1902, with the backing of a friend, he set up the American Safety Razor Company. (The use of his partner's name, Nickerson, was considered by both of them to be ill advised, given the nature of the business.)

The fact that Gillette had any steel to sharpen at all was itself the result of somebody else's bright idea over a century earlier. His name was Benjamin Huntsman and in spite of the fact that he wasn't a steelmaker, it was because of him that the Bowie knife and the best cutlery in the world were made in Sheffield, England.

In 1740, when Huntsman arrived at a village near Sheffield, steelmaking was a laborious and time-consuming process. Flat bars of iron were laid in a furnace chest, side by side on a bed of charcoal. The bars were then covered with charcoal, another layer of iron bars was placed on top of that and the process was repeated until the chest was full. It was then placed in the furnace, covered with a layer of sand and cooked red-hot for a week. During this time the carbon from the charcoal was absorbed into the outer layers of the iron, the carbonized areas forming blisters on the surface of the bars. When the chest was removed and had cooled down, these blisters were hammered off, and the pieces reheated and hammered together. The resultant "blister" steel was brittle and difficult to work.

Huntsman got his idea for how to solve the problem (to remelt the blister steel) from watching glassmakers at work. They would often reheat bits of old glass at very high temperatures to make it remelt and fuse. In order to be able to remelt blister steel, Huntsman invented a new clay container called a crucible, which was able to take the required 1,600°F (878°C) because of a mystery ingredient he added to the clay. The secret died with him, but it was probably black lead. Be that as it may, Huntsman's crucible steel was the first that could be cast. It also had high tensile strength, was so hard it would cut glass and would serve as the cutting edge of machine tools. The steel was so successful when Huntsman exported it to France that the Sheffield cutlery makers finally began to use it. The first cast-steel cutthroat razor was produced there in 1777, and a sales representative was in New York by 1800.

However, since Huntsman was a clockmaker, the feature of crucible steel that interested him untsman most was the fact that it would make a good clock spring. And — in one of those delightful links that history makes — in Huntsman's time public attention was focused on clock springs (made of the kind of steel that would also make a good razor), thanks to what happened to a British admiral who invented a wig. He rejoiced in the name of Sir Cloudesley 283 133 99 Shovell,[283] and the "full-bottomed" wig he designed (known as the Shovell wig) cost so much that only the rich and important could afford one. These people became known as bigwigs.

Alas, Shovell was only briefly to profit from his hirsute invention because in 1702 he drowned, together with the two thousand crewmen in his fleet of four ships. That foggy night, Sir Cloudesley was bringing them back from Gibraltar to England when, because of a major positional error, they hit the rocks off the Isles of Scilly and went to the bottom.

By 1714 Parliament had received a petition asking it to do something to improve maritime navigation, since an increasing number of explorers and colonists were being lost at sea, with alarming regularity. The upshot was that Parliament offered a massive prize (equal to about two million dollars today) for anybody who could come up with a better clock.

The importance of a timepiece was due to the fact that the calculation of east-west longitude involved having an onboard clock that would show exactly what time it was back at home port. When navigators took a star fix, the difference between the position of the star as they were observing it and where it would have been at that precise moment back home told them how far around the Earth they were. But since the Earth turns one degree in four minutes, and a degree of longitude is sixty miles, getting the time wrong by four minutes would mean being sixty miles off course. Since the horizon was only thirty miles away, this margin of error was a serious matter if the navigator was looking for landfall on an island.

284 266 224 In 1762 another clockmaker, John Harrison,[284] provided the answer because, like Huntsman, he was interested in the way metal behaves. Although Huntsman had developed an improved clock-spring steel that would not weaken over time and cause a clock to

slow down, this innovation ignored one particular difficulty encountered by world travelers. As a ship moved in and out of polar, temperate and tropical latitudes, the changes in temperature would cause the spring to expand or contract and, in this way, degrade the instrument's precision.

Harrison's idea was beautifully simple. He attached a tiny brass slider to the anchored end of the steel spring. Knowing the linear-expansion characteristics of both metals, he made the slider of precisely such a length that as the brass expanded and contracted, the slider moved up and down to immobilize longer or shorter lengths of spring. In this way, as the spring also expanded and contracted, the amount of spring left free by the brass slider to operate the watch always remained the same. On the final test run from London to Barbados, Harrison's chronometer ran accurately to the equivalent of a tenth of a second over fifteen months. So now, after a transatlantic voyage, a homecoming ship could hit the spot to within five hundred yards — fantastically accurate, but not accurate enough.

There is little point in returning home on a dark and stormy night to a position accurate to within five hundred yards if the treacherous rocks *also* within five hundred yards are not indicated by a lighthouse. And in the middle of the eighteenth century, such forms of illumination were all too infrequent. Lighthouses[285] had been erected around the coast of Europe since early in the previous century, but they were generally victims of their own success. If the illumination of a predominantly wooden lighthouse was particularly bright, the large number of candles involved tended, sooner or later, to burn the lighthouse down. Often these flimsy structures were also smashed by storms.

One of the most dangerous stretches of water round the British coast (partly because it was also so heavily frequented) was at the Eddystone rocks, fourteen miles out from the major British port of Plymouth. This location was famous both for lost ships and for lost lighthouses. By 1756 two lighthouses had been destroyed, and the local merchants were keen to replace them with one that would last. So they approached an engineer who had already invented a diving bell, redesigned major harbors, experimented with water-wheels and was generally familiar with things both maritime and hydraulic. His name was John Smeaton, and he designed a curved-profile Eddystone lighthouse made of dovetailed granite blocks and a new kind of cement that was the hardest ever seen, wet or dry. So

285 172 *135*

his lighthouse never fell down. It became so famous that when he helped to found the first Society of Civil Engineers in London, the lighthouse was on its crest.

Not much else of note happened to Smeaton, except that in 1770 he designed a machine for boring out water-pump cylinders. However, his invention was outdated by somebody else's four years later. The new borer was the work of England's greatest ironmaster, a weird type named John Wilkinson, whose sister married Joseph Priestley,[286] whose fizzy soda water kicked off this story. Wilkinson was obsessed with iron. He built an entire church made of it, paid his workers with his own iron money and had himself buried in a cast-iron coffin. At night, he would sleep with an iron ball in his hand. If he stirred (because he'd dreamed of a good idea), the ball would fall into a metal jug. The noise would awaken him, he'd make a note of the idea, pick up the ball and go back to sleep.

In 1774 he designed a cylinder-borer which used a rigid shaft with a cutting head on the end. The cylinder to be bored was rotated against the head. In this way, Wilkinson was able to cut cylinders to dimensions correct "within the thickness of an old shilling." This was the kind of precision that made it possible for James Watt to build his new steam engines and start the Industrial Revolution.

Wilkinson's borer had a revolutionary effect in another, equally meaningful, way. The machine would also bore out cannon barrels, and in 1788 Wilkinson achieved the dubious distinction of supplying the barrels to both sides in the Anglo-French war, as well as to the American colonists fighting England. Wilkinson managed to smuggle the guns to the French (disguised as "iron piping") because two years earlier he had supplied all the piping for the new Paris water-supply system and was an adviser to the French ironworks at Le Creusot. The new cannon barrels were thinner and weighed less than previous types. This minor improvement was to change the face of war (and Europe) because it enabled Jean-Baptiste Gribeauval, the French inspector general of artillery, to change the way guns were used in battle. In 1765 he had already begun to standardize the weight of French field artillery shot into four-, eight- and twelve-pounders. He had reduced the weight of the guns by removing all the ornamentation (previously so common) and standardized ammunition according to its intended range of fire. He also instituted a standard means of ammunition delivery to the troops. The point of standardization is that it made replacement and repair easier.

Wilkinson's new cannon barrels would change everything once they got into the hands of an ex–artillery lieutenant, now master of France, Napoleon Bonaparte.[287] Gribeauval had profited from the lightweight, interchangeable, precision-made Wilkinson cannon barrels to create an entirely new, mobile artillery. After 1799 the crack French unit was Napoleon's nine-regiment Horse Artillery of the Guard, about whom he said, "It is the Artillery of my Guard which decides most of my battles, because I am able to bring it into action where and when I wish." Thanks to Wilkinson's new, lighter barrels, the guard could do something entirely new in warfare. It could rapidly move up to forty cannons around the battlefield, as desired, and split them in two or more groups, so as to pound the enemy with crossfire and soften them up for the infantry.

In Napoleon's case, this was a vital tactic since his infantry was generally composed of hundreds of thousands of raw, untrained conscripts who would have been no match for professional enemy soldiers — unless the latter had already been dismembered by artillery. As Napoleon said, "The less good troops are, the more artillery they require. There are *corps d'armée* with which I should require only one-third of the artillery which would be necessary with others."

As he developed his tactics, Napoleon used the powerful destructive force of massed artillery to greater and greater effect. The aim of this was brutal: "The object of artillery is not to kill men or dismount pieces in isolation, but to make holes in the enemy front, stop his attacks, and support those launched against him." At the height of his success, Napoleon was using the horse artillery's advantage of speed and surprise to mass as many as a hundred guns in one place, against which "nothing will resist, whereas the same number of cannons spread out along the line would not give the same results."

Napoleon ranged across Europe, invincible. In 1799, when marching through Switzerland on their way to fight the Austrians, his victorious armies encountered armed opposition in the canton of Unterwalden. In the ensuing fight, hundreds were killed, and their children orphaned. In Stans, the canton capital, a middle-age ex-farmer decided to take some of the homeless children into his care, at a converted ex-monastery. So began one of the most extraordinary and important educational experiments in history. Johann Heinrich Pestalozzi had failed in agriculture and also as a

287 96 66

novelist. But the orphanage at Stans was to make him famous all over the world. It may have been because there was no money for books and equipment — or that it was a time when schooling was a luxury only the well-off could afford — which led Pestalozzi to formulate his plans for a new kind of teaching. In 1801 he put the idea in writing with a textbook called *How Gertrude Teaches Her Children: An Attempt to Give Directions to Mothers How to Instruct Their Own Children.*

Pestalozzi's theories were radically different from anything that had gone before. In his view, the education of the time was based on academic book-learning and took no account of the development of the whole child or of the relationship between lessons and the real world. Children were usually told of mountains without ever having climbed a hill and of duty and virtue without having the faintest idea what the words meant. Pestalozzi's technique was aimed at developing the child's powers of perception and intuition, through learning of the world by direct experience. He was against all formality, so he did not divide the children into classes or use any books or other school materials. He encouraged students and teachers to behave as equals, and he permitted children to teach other children. Pupils and teachers often slept in the same room and even studied together. The daily routine began with the first lesson at 6 A.M., morning prayer at 7 A.M., then breakfast, followed by more lessons beginning at 8 A.M. A snack at 10 A.M. ended with an hour's recreation and more lessons. Lunch was at 1 P.M., lessons continued till 4:30 P.M., recreation till 6 P.M., classes till 7 P.M., supper at 8 P.M. and bed at 10 P.M.

Subjects were taught with reference to their real-life context. Geography class involved observation during field trips, after which the children drew a map that included details of what they had seen. Only then were they shown a real map. Science was similarly based on the elements of their local environment, for instance, observing how wine turned to vinegar, sand into glass, marble into lime and so on. Music was employed as an aid to moral education and to develop a sense of social harmony. Games, played twice a week, were intended to develop a sense of team spirit. Above all, the children were encouraged to learn from their own daily lives, during frequent walks in the open air, and to use this experience to make sense of the world in their own way.

Pestalozzi's method achieved almost overnight acceptance. What had begun as a way of rehabilitating children traumatized by war would one day inspire educators all over the world. By 1805 the method was beginning to be used throughout Europe, and there was already a Pestalozzi school in Philadelphia. At one point, his techniques would be a key part of the famous (and briefly successful) socialist community in New Harmony, Indiana, thanks to their effect on William Maclure,[288] one of the community's founders.

288 258 *216*

Pestalozzi's emphasis on using direct experience to help a child move from "confused intuition to clear perception" endeared him to a German academic he had met during a visit to Interlaken, Switzerland, in 1797. Johann Friedrich Herbart was to become a close friend and one of Pestalozzi's most effective advocates. By 1808 Herbart was Kant's[289] successor in the chair of philosophy at Königsberg University. A recognized authority on education, Herbart introduced

289 210 *170*

Pestalozzi's ideas into the academic mainstream by writing a report on his methods in a work called *Pestalozzi's ABC of Observation.*

Herbart built on Pestalozzi's ideas of sense perception and his natural method of teaching and, in doing so, developed the first proper scientific pedagogy. He recognized the social implications of this new approach to education: "Pestalozzi's essential aim has been to raise the lower classes and clear away all differences between them and the educated classes. It is not only popular education that is thus realized, but national education. Pestalozzi's system is powerful enough to help nations, and the whole human race, to rise from the miserable state in which they have been wallowing."

Herbart took Pestalozzi's ideas a stage further by looking at what actually happens in the individual whose perceptions are involved in the process of learning from experience. Herbart's idea was that each experience modifies a person, so that the recognition and interpretation of new experiences is affected by the influence of earlier experiences, which are already a part of the personality. This accumulation of experience becomes what Herbart called the "apperceptive mass," the totality of experience with which the individual understands the world. As part of this cumulative process, any experience immediately becomes at one with the mass of earlier but similar experiences. But if the experience is new, it crosses what Herbart called the "threshold of consciousness" (a term he coined) to become consciously identified as a "new" experience. Herbart thus established the formative nature of this threshold and effectively turned psychology into a science.

The big issue now was to discover at what point an experience crosses the threshold and becomes a "conscious" experience. Could this point be identified? Could the process be quantified? In 1833 Herbart moved to the University of Göttingen, where one of his new teaching colleagues was a man named Ernst Weber, who had a pupil, Gustav Theodor Fechner. Between them, Weber and Fechner found a way to measure the effect of experience. Weber produced a formula, and Fechner applied it to "measuring the mind" and invented what he called "psychophysics."

Fechner was a strange man. His first writings were satirical, with titles like *Proof That the Moon Is Made of Iodine* and *The Comparative Anatomy of Angels* (in which he argued that since angels were supposed to be perfect, they must be spherical). In 1839 a serious illness which resulted in three years of partial blindness (brought on by experiments in perception that involved gazing at the sun through colored glasses) forced him to stop teaching. In 1843, when he recovered, the experience of his first sight of flowers was so powerful that he wrote a book called *Nanna, or the Soul Life of Plants*. Fechner also became very interested in Persian mysticism and invented "pan-psychism," in which he placed the human soul between the souls of plants and those of stars.

Then, in 1850, he got down to the work of quantifying experience, which would end with his formulation of the Law of the Just-Noticeable Difference. In spite of its name, this law was to become a matter of cosmic significance. Using Weber's data, Fechner conducted a long series of experiments to find out what causes an experience to cross Herbart's threshold of consciousness and become noticed. Fechner believed that this event is related to the level of stimulus involved in the experience. Crossing the threshold of consciousness seems to have something to do with the extent to which any stimulus is greater or lesser than a previous one. In other words, Fechner was trying to find the smallest change in stimulus that would cause an individual to notice the difference.

Fechner and Weber wanted to show that sensation and stimulus are mathematically interdependent, so that one can be predicted from the other. Their experiments consisted of gradually decreasing the difference between two stimuli until the difference became indiscernible. They tried this with different weights, with the judgment of lines of different lengths, and with sounds, light intensities, smell, temperature, the pitch of notes and so on.

What they discovered was that there is indeed a constant factor at work in perception. If the initial stimulus is doubled, any addition also has to be doubled in order to be noticed. If the least noticeable extra weight added to a weight of fifty pounds is one pound, then the least noticeable extra weight added to a weight of one hundred pounds is two pounds.

The least noticeable difference is, therefore, a constant which is proportional to the original stimulus. This became a mathematically definable law relating to sensory magnitude, which could be applied to all sorts of experience in a quantifiable way. With it, Fechner effectively began the modern science of experimental psychology.

One of Fechner's other ideas involved the luminosity of stars. He thought his new law explained the scale of perceived stellar magnitudes used in astronomy. Stars, he said, are invisible during the day because the difference between their brightness and that of the daytime sky is too small to be noticeable. Earlier in the century, a Munich professor of physics called Carl Steinheil had invented a device for measuring stellar brightness. Called a photometer, it was also used to calibrate gaslight[290] flames. It consisted of a telescope in   290 107 73 which the object glass was split into two halves, one of which was moveable in the direction of its axis. Images of two stars could therefore be brought together by the two half-glasses, and then a lens could be used to change the image of one until it was perceived to be as bright as the other. The amount by which the lens had been moved would provide the data by which the relative brightness of the two could be identified.

Using the photometer, Fechner quantified the just-noticeable difference — which had until then characterized the astronomical definition of different stellar magnitudes — to be about 2.5. In other words, a star of first magnitude was about 2.5 times brighter than a star of second magnitude. A star of first magnitude was exactly 100 times brighter than a star of sixth magnitude. This was to prove essential in judging how far away stars are. It was already possible to spectrally analyze starlight to show the elements burning in the star, which indicated their approximate temperature and, consequently, their real brightness.

In terms of magnitude of brightness, the brilliance of a star diminishes as the square of its distance (if a star is twice as far away, it will be one-fourth as bright); these figures give accurate estimates of stellar distance. Using these and other techniques, in 1908, Henrietta Leavitt, an astronomer at Harvard College Observatory, studied certain unusual stars called Cepheids. Their brightness waxes and wanes, and Ms. Leavitt discovered that this change in luminosity varies with their period. In other words, the length of time between brightest and darkest luminosity indicates the true brightness of the stars. From this and spectral analysis, it is possible to work out their distance from the Earth.

This discovery regarding Cepheid brightness turned out to be of extraordinary significance. On October 5, 1923, another American astronomer, Edwin Hubble, working at Mount Wilson Observatory in California, found Cepheids in the outer reaches of Messier 31, the great starcloud in the constellation of Andromeda. The distance of the Cepheids showed that Andromeda lay outside the Milky Way. Further calculation revealed that they were over 750,000 light-years away. This distance is over 600,000 light-years outside our own galaxy. The figure caused astronomers to double the previously accepted size of the universe.

So thanks to bright ideas like bottle caps, razor blades, chronometers, cannon borers, new teaching methods, experimental psychology and stellar magnitudes, in 1934, Hubble changed our view of the universe forever.

In an electrifying lecture, he said, "The stellar system is drifting through space as a swarm of bees through the summer air. From our position somewhere within the system, we look out through a swarm of stars, past the borders, into the universe beyond . . . here and there, at immense intervals, we find other stellar systems comparable with our own . . . so distant that, except in our nearest neighbors, we do not see the individual stars of which they are composed."

Hubble's later discovery — that this immense cosmos is also expanding — eventually led to new theories about the nature of the universe, a matter about which humankind has been philosophizing since the Middle Ages. . . .

19 Echoes of the Past

S OMETIMES, as this book has shown, the cause of major change is something minor — like accident, luck, individual greed, ambition, a mistake or some other such incidental matter — that gives a sudden twist to the path of history. In this case, however, one of humankind's most important discoveries about the universe was triggered by two events, widely separated in space and time, both of which were actually concerned with nothing less than how the cosmos worked. The second of these events began with developments in cryptography, in early fifteenth-century Florence.

The first was the arrival of Zen Buddhism in medieval Japan. In 1133 a monk called Eisai Myo-an went from Japan to China and returned with tea seeds, which he planted around a temple in Hakata, Kyushu. There had been tea-drinking ceremonies in China since the eighth century, and there were several reasons for bringing the ritual to Japan — one of which was that tea was considered to have great medicinal value, being regarded as the remedy for all disorders including beriberi, paralysis, boils and loss of appetite.

But it was the link between tea drinking and philosophy which made its mark in shogunate Japan, dominated at the time by the samurai warrior ethic. The tea ceremony was an expression of the Zen Buddhist view of life, and Zen appealed to the practical-minded soldier because it is a religion of the will; for the true believer, as for soldiers, life and death are matters of little importance. Like a military commander, a Zen Buddhist never looks backward once a course is set or a decision taken. Zen also appeals to the intuition, rather than the intellect. And there are no complex liturgical formulae to be learned.

The tea ceremony itself, which lasted over four hours, appealed to the instinct for discipline. Its rigid rules of etiquette formed a bond of ethics between host and guest. The basic elements were extremely simple. Rules were precise about how to make the tea, how to serve and drink it, what utensils to use and how to whisk the tea with fresh-split bamboo. The cup was to be placed, picked up and returned to a tatami mat in strict sequence, with the correct hand movements and gestures of appreciation by the guest. The cup should be picked up with the right hand and placed on the left palm, with the fingers of the right hand round it, thumb facing inward. As a guest did this, he or she should give a small bow to the host. With the cup still resting on the left palm, the guest should grasp its rim with the forefinger and thumb of the left hand, turn it ninety degrees clockwise, sip once and comment on how good it tasted. At the same time the right hand should rest on the tatami mat by the guest's knees. Several small sips should then be followed by a final deep and audible sip.

Then the cup rim should be lightly wiped from left to right with tissue, while it was still held in the left hand. At this point, the cup should be held in the left palm. Using the fingers and thumb of the same hand, the guest should turn the cup counterclockwise to its original position, with thumb at the rim and four fingers below it. The cup should then be placed down, outside the edge of the tatami mat. The guest should then leave his or her hands resting on the mat, gaze at the cup and return it to the host, first turning it so that its front faced him or her. All these rules applied only to the drinking of thin tea. Rules for thick tea were different.

By 1610, when the Dutch explorers arrived in Japan, the tea ceremony was a long-established upper-class ritual, and every imperial prince had his own personal tea adviser to help choose among sixty different types of brew. The ceremony was also now accompanied by the consideration of philosophical questions. Such questions as "What is the nature of the universe?" and "What is the sound of one hand clapping?" were intended to help the questioner discover the Infinite Oneness of Everything. Tea drinking was an essential step on the Way to Illumination.

To the astute Dutch merchants, it was also a step on the way to profit. When China tea (bought in Japan, because the Chinese would not permit direct trading) first arrived in Holland, it created a sensation. By 1640 it was being drunk by all the social classes in bizarre, afternoon versions of the tea ceremony, in which Dutch middle-class etiquette replaced Buddhist rituals. The tea was accompanied by cakes, brandy, raisins and pipes of tobacco. Tea drinking became a craze; by the end of the seventeenth century, three-quarters of all trade with the East was in tea, and the beverage was an indispensable luxury for the European bourgeoisie.

But the real profit to be made in the tea trade was to be made not from the tea, but from the porcelain[291] cups in which it was served. In 1602 the Dutch had captured a Portuguese carrack, the *Santiago*, carrying a cargo of porcelain. The pieces that went on sale at Middelburg netted a colossal profit. By 1637 the Dutch East India Company was importing twenty-five thousand teacups a year, and by 1657 it had brought back over three million pieces. Demand was insatiable, and prices were sky-high because porcelain had become a royal collectible. The elector of Saxony (who was also king of Poland), Augustus the Strong, set the style with special rooms built to house his collection. It was he who would help the Europeans find a way to make porcelain almost as well as the Chinese did.

In 1682 he permitted a refugee named Johann Böttger to settle in Magdeburg, and in 1708 Böttger succeeded in producing a fine, white transparent ware. Although it was not the real thing, it did well enough. By 1709 Böttger had moved to Holland and was working in the port of Delft, where there had been a pottery industry for some time. The style of the new delftware Böttger created was a copy of the porcelain brought back from China. At the time (the Kangxi period), the fashion in China was for blue-and-white floral motifs, or landscapes with clouds and figures, placed within rectangular frames. These designs set the classic porcelain style[292] in Europe from then on, as did the contemporary Chinese globular or eight-sided teapots, square tea caddies and cups without handles. There were also decorative pieces called chargers, large flat plates which were often hung on the wall.

Delftware was soon so much in demand and so valuable that it

291 11 *18*

291 243 *205*

292 124 *94*

293 223 *185* was cheaper to have a broken piece repaired than replaced. In 1759
293 307 *278* the English potter Josiah Wedgwood[293] went into business fixing
broken porcelain and delftware. By then a few English potters, including Wedgwood, were using a new clay from the English coun-
ties of Dorset and Devon and were producing a cream-colored
ware. In 1762 Wedgwood met and befriended Thomas Bentley, a
Liverpool merchant and intellectual (he was a friend of Joseph
Priestley's) who had just done the grand tour of Europe. The tour
was fashionable among rich young men, who would visit the culture
capitals of Europe, accompanied by a personal tutor. During the
trip they would often pick up souvenirs, including paintings, statues
and other "collectibles" to take back home. Some of these collections
were so big that museums were created to house them.

Wedgwood and Bentley became partners, and when somebody
gave Wedgwood a copy of a new book by Sir William Hamilton, it
changed Wedgwood's life. The book illustrated Hamilton's collection of vases, wrongly identified by him as Etruscan. Wedgwood
copied the designs, produced a new kind of creamware china and
presented a set to the royal family. When Queen Charlotte expressed pleasure at the gift, he promptly named the design Queens-
ware. The new style became a runaway success all over Europe, and
by 1774 it would be so fashionable that Catherine the Great of Russia would order a 952-piece set, known from its motif as the
"frog" service.

Wedgwood's new patterns (now including blue on white, and red
on black, as well as the outstandingly successful white on pale blue)
were based on his recently acquired pictures of classic Roman and
Greek art. The fashion for things classical had been triggered by
the stupefying discovery in 1730 of the buried city of Pompeii. By
the late eighteenth century, books entitled *Ruins of* . . . were being
published all over the Continent. The archeological discoveries put
an end to the delftware fashion for *chinoiserie.* Neoclassicism rapidly
swept Europe off its feet, thanks principally to the work of Giam-
battista Piranesi, a Venetian draftsman who, in 1743, published his
first prints of Roman ruins. In a sequence of illustrated books, Piranesi's magnificent (and often fanciful) drawings of massive, de-
caying temples and basilicas fired the imagination of every designer,
historian and architect in Europe. One of these was a twenty-nine-
294 224 *185* year-old Scot, Robert Adam,[294] who met Piranesi when he was on

the grand tour in 1757 and accompanied him on expeditions to Hadrian's villa and along the Appian Way. Adam went on to do his own sketching in Rome and, later, at the Palace of Diocletian at Split, in Croatia.

When he returned to England, Adam introduced the new style to the smart world of George III. All those with pretensions to taste were soon commissioning Adam to build or redesign their houses. In them, he re-created the world of ancient Rome and Greece, with coffered ceilings, colonnades, porticoes and half-columns. He decorated interiors with pedestals, urns, medallions, trophies and patrons' insignia (most of them made for Adam by Wedgwood) and filled the rooms with pseudo-Roman furniture made by Thomas Chippendale.

But it was something else Piranesi had included in his *Views of Rome* that appealed to the Freemasons, an organization of liberal thinkers dedicated to worthy social causes and whose membership included people as diverse as Jefferson,[295] Voltaire, Goethe, Franklin, Washington and Mozart.[296] What caused the furor was Piranesi's 1769 *Different Designs for Fireplaces*, which included illustrations of ancient Egyptian motifs such as obelisks, sphinxes, pyramids and such. These caught the imagination of Freemasons, whose first grand lodge had opened in London in 1717. The members were fascinated by things ancient (they traced their origins to the classical world), so they promptly modeled their rituals and codes on Egyptian antiquities.

295 80 *53*
295 113 *83*
295 213 *172*
296 188 *151*

Freemasons were particularly influential in France, where under Louis XVI there was a positive frenzy of Egyptomania, as a cursory glance at the number of French towns with obelisks and sphinxes dating from that period shows. The most notable Freemason in the world at the end of the eighteenth century was, of course, Napoleon. His entire family and most of his relations belonged to the order, as did most of the French upper class, a fact which was said to have helped hold the empire together. When, in 1798, it became expedient for Napoleon to do something about the English threat, and he chose to cut off their link with India by invading Egypt, for a Freemason the cultural opportunity was too good to miss. On the one hand he would make Egypt a French colony, repair and modernize its crumbling infrastructure and build a Suez Canal (a plan later abandoned when his engineers told him the Red Sea was thirty inches higher than the Mediterranean). On the other hand, he would order the cataloging of this immense repository of ancient civilization by a team of experts who would decide which pieces of it to remove for exhibition in the Louvre.

In order to complete this great work of research, Napoleon appointed a commission of eminent scientists. He based the decision to do so on the fact that his role model, Alexander the Great, had always traveled with his own group of savants, with whom he would indulge in intellectual discussions whenever he wasn't making war. Napoleon was in any case extremely interested in science (partly because of his military background in artillery) and appointed a mathematician and a chemist to ministerial posts in his government. He also offered prizes for innovations that would put French science in the vanguard.

As soon as he arrived in Cairo, Napoleon set up the Egyptian Institute and gave it an agenda. He considered the work of the institute so important that while in Egypt he always signed himself first "Member of the Institute," and only then "Commander in Chief." In the arts, the institute's brief was to study monuments and antiquities, write a history of ancient Egypt and prepare a Franco-Egyptian dictionary. Institute engineers were also to devise schemes for freshwater reservoirs and to discover the causes of annual flooding. Agricultural experts were to experiment with new crops. Doctors would investigate the prevalent disease of ophthalmia, as well as reorganize the country's sanitary and hygienic systems. Chemists would analyze Egypt's potential for the production of gunpowder and dyestuffs. In addition, the scientific team studied desert mirages, crocodiles and hippos. A census was planned, the country was to be accurately mapped, and its geology and natural history chronicled.

The commission members were drawn from the widest circle of French intellectual life and included chemists, zoologists, engineers, mineralogists, physicists, economists, metallurgists, artists, doctors, archeologists, poets and playwrights. One of them realized the importance of a stone, uncovered by a soldier at the village of Rosetta and inscribed in three ancient languages: two of them differing versions of Egyptian script and the third, Greek. The stone was taken to nearby Alexandria (where it was later surrendered to the English after the French defeat), and impressions of the inscriptions were made. These would ultimately be used by the French scholar Jean François Champollion, in 1822, to decipher the stone's mysterious hieroglyphs.

Meanwhile, other secrets were to be recovered from beneath the surface by another member of the commission. He was a mathematician named Jean-Baptiste Fourier, whom Napoleon appointed permanent secretary to the Egyptian Institute and whose job was to investigate the monuments of upper Egypt. Fourier was a favorite of the emperor, who made him prefect of the province of Isère and a member of the Legion of Honor on his return to France. While in Egypt, Fourier contracted a thyroid disease, possibly myxedema, which brought about thickening of the lips and tongue, as well as loss of hair and memory. Above all, it caused him to suffer from the cold for the rest of his life. It may have been the need to spend most of his time in well-warmed rooms that caused Fourier to turn his attention to heat.

In 1807 he submitted to the French Institute the first truly major discovery in physics since the experiments of Newton. His first paper, on the diffusion of heat, gave the first mathematical description of how heat acts. Fourier defined the properties that cause bodies to contain, transmit or receive, and conduct heat. Each of these three basic properties is affected by the physical characteristics of the body. As one of the applications of his theory, Fourier selected the variations in terrestrial temperatures. He noticed that while the Earth's surface absorption of solar heat varies with the time of day and the season, temperatures underground remain constant. Fourier tried to arrive at a law which would explain how underground temperatures might be affected by variations in heat at the surface. His surprise discovery was that temperatures are actually higher, deeper down.

Some years later Fourier's heat-diffusion theory was used to calculate the age of the Earth by a Scotsman, William Thomson. His work showed that Fourier's heat-transfer laws mean that, as heat increases with depth, the planet must have originated in a molten state. And as diffusion of heat gradually lessened over time, the Earth must have cooled down much more rapidly at first. If this had been the case, in the ancient past, when the planet was losing heat from the molten core through volcanic action, the surface should have been much hotter than today. This view appeared to be vindicated when archeologists and geologists discovered fossilized palm trees beneath Paris.

It was also William Thomson who was to take Fourier's work to an extraordinary conclusion, when he put it together with the research of another Frenchman, an engineer called Sadi Carnot. Carnot, who was a generation younger than Fourier, was the son of Napoleon's minister of war and believed that Napoleon had been defeated by the British because of the tremendous industrial production made possible by their steam engine. At Waterloo, for instance, Napoleon's army had been firing guns and wearing uniforms made in Britain. In Carnot's view, only the discovery of some way to build a superefficient steam engine would restore French glory.

In 1824, after studying steam power in Paris factories and workshops, Carnot wrote a paper on the "power" of heat. By this, he meant the way in which heat causes motion in steam engines. His theory was that this power comes from the one-way-only transfer of heat, from a hotter to a colder body, and that during this transfer, heat can be made to do work when it goes from a higher to a lower temperature. Within a couple of decades William Thomson (who was later to become Lord Kelvin) modified and improved Carnot's theories, which became a general theory of heat.

In 1851, while he was professor of natural philosophy (physics) at the University of Glasgow, Kelvin came to the conclusion that there might be a temperature below which no further work could be produced. Since heat is produced by moving molecules, the temperature should fall no farther at the point where all molecular motion in a gas stops; therefore, no more work could be produced below that temperature. He calculated this final chill to be $-273°C$ ($-459.4°F$), calling it absolute zero on a temperature scale he developed, the Kelvin scale. On this scale, each degree is equivalent to a degree on the Celsius thermometer.

Pulling together all that was known about heat transfer and diffusion, Kelvin worked out the second law of thermodynamics: without added energy input, everything hot will become cold, and all structures will eventually break down and become formless. Therefore, this law predicts that the universe should have gradually cooled down from an initial hot big bang.

The discovery of the second law of thermodynamics, which describes the behavior of the cosmos, ends the first of the two histori-

cal trails with which this chapter begins: one that started with medieval Zen Buddhist tea drinkers, who were — ironically — seeking
an understanding of how the universe worked. The chapter's second
trail (which ends with a discovery that could only be made thanks
to Kelvin's thermodynamic laws) begins with another search for the
secret of the universe, this time in Renaissance Italy.

In the early fifteenth century, polymath Giovanni Battista Alberti wrote *On Architecture*, a major treatise based on newly discovered classical texts. In it, he set out the new rules for the use of
perspective geometry in building. His own work in Florence, like
the facade of the church of Santa Maria Novella, and the church
of San Lorenzo, are beautiful examples of the general Renaissance
concern for *eurythmia*, or balance, in everything. This universal harmony was believed to originate in the "music of the spheres," the
mystical heavenly tones that were supposed to echo through the
universe and that were a manifestation of the essential unity beyond
God's Creation.

In architecture and other forms of art, *eurythmia* was thought to
exist primarily in numbers. So Alberti's work on Santa Maria, for
instance, is a giant mathematical game. The facade can be divided
into rectangles or squares. Each one of them is in ratio to some
other part, and all of them are in proportion to the whole. Geomet-
ric relationships like these — that could be generated with numbers — were supposed to have a mystical quality, and part of the
general Renaissance preoccupation with them also showed itself in
an obsession with puzzles and codes. Alberti was more involved
with this than anybody else was at the time, and he wrote a book
on cryptography that would influence the world of spies for over a
hundred years. In the fifteenth century the reason for this intense
interest in secrecy was due to the growing number of nation-states
and their burgeoning diplomatic corps. Increasing numbers of am-
bassadors and envoys (and especially those of the Vatican) were
sending increasing numbers of secret, politically sensitive messages
to one another.

By 1585 a Frenchman named Vigenère, who was a fan of Alberti's work, had produced a code that he claimed nobody could
crack. It was to become the most widely used cryptographic system in Europe. The code was based on a rectangular matrix of
letters:

```
A B C D E F G H I J K L M N O P Q R S T U V W X Y Z
B C D E F G H I J K L M N O P Q R S T U V W X Y Z A
C D E F G H I J K L M N O P Q R S T U V W X Y Z A B
D E F G H I J K L M N O P Q R S T U V W X Y Z A B C
E F G H I J K L M N O P Q R S T U V W X Y Z A B C D
F G H I J K L M N O P Q R S T U V W X Y Z A B C D E
G H I J K L M N O P Q R S T U V W X Y Z A B C D E F
H I J K L M N O P Q R S T U V W X Y Z A B C D E F G
I J K L M N O P Q R S T U V W X Y Z A B C D E F G H
J K L M N O P Q R S T U V W X Y Z A B C D E F G H I
K L M N O P Q R S T U V W X Y Z A B C D E F G H I J
L M N O P Q R S T U V W X Y Z A B C D E F G H I J K
M N O P Q R S T U V W X Y Z A B C D E F G H I J K L
N O P Q R S T U V W X Y Z A B C D E F G H I J K L M
O P Q R S T U V W X Y Z A B C D E F G H I J K L M N
P Q R S T U V W X Y Z A B C D E F G H I J K L M N O
Q R S T U V W X Y Z A B C D E F G H I J K L M N O P
R S T U V W X Y Z A B C D E F G H I J K L M N O P Q
S T U V W X Y Z A B C D E F G H I J K L M N O P Q R
T U V W X Y Z A B C D E F G H I J K L M N O P Q R S
U V W X Y Z A B C D E F G H I J K L M N O P Q R S T
V W X Y Z A B C D E F G H I J K L M N O P Q R S T U
W X Y Z A B C D E F G H I J K L M N O P Q R S T U V
X Y Z A B C D E F G H I J K L M N O P Q R S T U V W
Y Z A B C D E F G H I J K L M N O P Q R S T U V W X
Z A B C D E F G H I J K L M N O P Q R S T U V W X Y
```

To use the matrix, a code word was agreed upon by the sender and recipient (say, "Booth"). First, the message (say, "Lincoln Dead") was written under the code word (which was repeated as many times as needed:

BOOTHBOOTHB

LINCOLNDEAD

To encode "Lincoln Dead," decoders found the first letter, *L* (on the top line) and the row that begins with the letter *B* (second row). Where the column and row intersected was the first code letter, *M*. This process continued until all letters of the message were encoded as "MWBVVMB RXHE." To decode the message, the recipient found the first code-word letter, *B*, along the top row of the matrix, went down that column to find the first code letter, *M*, and then back along that row to the first letter in the row (to the left side of the matrix), where the decoded letter was revealed as *L*. This process was repeated for the rest of the coded message.

Had events gone differently, this particular message ("Lincoln Dead") might well have been the last recorded use of the Vigenère code during the American Civil War. The code was being used by the Confederates, and a copy of the matrix was found in John Wilkes Booth's hotel room when it was searched after his arrest for the assassination of Abraham Lincoln.

In fact, one of the reasons the South lost the war was that the North had cracked the code, and Washington was getting Confederate war plans within twenty-four hours of their formulation. Another major factor in the South's defeat was the ability of the Union commanders to feed and supply their troops and horses by rail. The railroads[297] were themselves part of the problem that had caused the war in the first place, because the northern railroad companies had diverted valuable western trade away from the southern ports to those of the Northeast and started carrying European immigrants to the midwestern corn belt.

Beginning in 1862 the Homestead Act[298] made land available at $1.25 an acre and attracted hundreds of thousands of would-be farmers. Apart from the need to supply the Union army, the growing eastern cities were already providing huge markets for any food the Midwest could produce. Because of the war and a consequent shortage of labor, farmers were encouraged to mechanize. They did so, and in the 1860s their productivity increased by 13 percent, thanks to the new McCormick reaper. By 1858 there were already seventy-three thousand of them at work, and 70 percent of western corn was being cut by machine. And as the railroads pushed west beyond the Mississippi, grain shipments rose by a staggering amount. In 1838 Chicago had shipped seventy-eight bushels, but by 1860 the figure was over thirty-one million.

In the second half of the nineteenth century, a tremendous boost in food production had been achieved thanks to mechanization, the Homestead Act, the demands of the Civil War and the development of immensely long, sixty-car freight trains. These could carry enough wheat to provide a year's supply of bread for ten thousand people. Reapers and freight trains and grain elevators all turned America into the world's greatest agricultural exporter. By the turn of the century, she was providing 29 percent of the world's wheat.

297 27 *28*

298 217 *180*

This had an immediate effect on the size and shape of ships, as American food began to cross the Atlantic in large amounts on board bigger and bigger bulk carriers. Although business preferred the more economical sailing ship, the attraction of regular steam service was soon irresistible. The most serious drawback to the new, larger cargo steamships was that they used tremendous amounts of coal. Initially, this problem was solved by John Elder and Charles Randolph, in 1855, with a system that fed the steam from the boilers into a small, high-pressure cylinder, where it drove the piston. The steam was then reused in a larger, low-pressure cylinder. So the ship's owners were getting double service from the same steam. These two-cylinder engines were in transatlantic ships by 1871. Then, in 1882, a third cylinder was added, and on the first major trial of this system, the *Aberdeen* completed her run from Britain to Australia with coal to spare.

In 1884 came another breakthrough. Charles Parsons invented an engine in which a jet of steam turned a multiblade turbine shaft that was connected directly to the ship's propeller. This innovation did away with the need for pistons at all. In 1897, at the annual Spithead review of the British fleet, the first ship fitted with the new engine, the experimental *Turbinia*, raced past the line of warships at 34½ knots, easily outrunning the torpedo boats that had been sent to intercept her. The Parsons turbine engine was so efficient it had to be geared down to speeds that would avoid loss of thrust through cavitation, which occurs when the propellers are going so fast that they literally make a hole in the water, so there's nothing to push against. The new turbine engines used fuel so efficiently that by 1910 they were in the first cargo ship, the British *Cairncross*.

Meanwhile, as cargo ships grew in size and speed, so too did passenger liners. In the mid–nineteenth century, the number of European emigrants crossing the Atlantic to America (seven million by 1875) made this by far the most profitable passenger route. However, by the 1870s the traffic was starting to pay in both directions, with American businessmen and tourists filling the liners on their return journeys. So the liners grew in size (thanks to the better engines) and comfort (thanks to innovation in general). In 1876 electric navigation lights were fitted to the French line's *Amerique*. In 1833 the *Normandie* got interior plumbing. In 1899 the *Oceanic* had marble lavatories and looked like a floating hotel. The White

Star's *Celtic*, the first vessel to exceed twenty thousand tons, was the last word in efficiency and comfort and carried three classes of passengers. Liners went on getting bigger and faster. In 1912 the America Hamburg Line's *Imperator* grossed 51,969 tons, and in 1929 the *Bremen* broke all speed records on the Atlantic with an average of 27.92 knots.

In 1927 the biggest ship in the world, the fifty-five-thousand-ton *Leviathan*, carried a radio telephone,[299] the ultimate in luxury for people who couldn't bring themselves to lose touch with their businesses. Just after leaving England, a passenger could place a three-minute call to New York for only $75. The only difficulty with this arrangement was that very often they couldn't hear the party on the other end of the line. For some mysterious reason, shortwave radio-telephones suffered from a great deal of interference. So in 1931 a Bell Telephone Laboratories engineer, Karl Jansky, was asked to investigate the cause of the static. Working with a steerable antenna, which was mounted on four Ford Model T wheels and automatically rotated every twenty minutes, Jansky found that the source of the static seemed to drift gradually east to west across the sky, throughout the day. But the strange thing was that it rose from the eastern horizon four minutes earlier each day until, after a year, it was appearing a full day earlier. At first Jansky thought the static might be due to shifting thunderstorm activity, but on investigation this proved groundless.

Then Jansky's astronomer friend and bridge partner, Melvin Skellett, remarked that because the Earth orbits the Sun, as the planet turns, stars rise four minutes earlier each day. Over a year, this change in star-rise time would amount to a full day. By 1932 Jansky knew the source of his static was extraterrestrial, coming from stars beyond the solar system. The following year, when he announced that the source of radio interference was the Milky Way galaxy, he became a world celebrity. Then, in 1965, two more Bell Labs researchers, Arno Penzias and Robert Wilson, made the great leap when they found that Jansky's static was not coming from the galaxy, but from all over the universe.

It is this momentous discovery which brings together the two strands of this historical tale that begins with medieval Japanese tea drinkers and Renaissance Florentine architects, both searching in their different ways for the secret of how the universe works.

299   15   22
299   236   198

Thomson's law of thermodynamics had shown that without added energetic input, everything hot became cold. He had also invented the Kelvin scale on which he placed absolute zero, the final end of all cooling processes, at $-273°C$ $(-459.4°F)$. By the time Penzias and Wilson were working, it was known that hot objects emit radiation and that this radiation can be detected as static on a radio receiver. Investigation of the whole-sky static they discovered in 1965 revealed that the temperature of the source of the static (the cosmos) is about 3.5° Kelvin. If the universe has indeed been cooling down since an initial, incandescently hot big bang more than ten billion years ago, then according to the second law of thermodynamics, the cosmic temperature should be just about 3.5° Kelvin.

The view of what (or who) is responsible for the Beginning of Things has been the cause of dissension and argument throughout history, rarely with more world-altering results than when the row surfaced in the late Roman Empire. . . .

20 One Word

L IFE in sixth-century Europe must have seemed a little like the post–Cold War era in the late twentieth century. The comfortable balance was gone. Pockets of imperial civilization still existed here and there, but in general the Continent was fragmented among barbarians who adapted, more or less successfully, to the lifestyle of the Romans they had conquered. In Spain a Romano-Hispanic majority was ruled by a Visigothic warrior elite that had been moved west from what is now Romania by the imperial Roman government, almost two hundred years earlier.

The Visigoths were the most Romanized of barbarians, and in A.D. 589 they controlled *the* largest and most powerful of the many kingdoms into which Europe had broken up after the fall of Rome. To the pope the Visigoths represented a potential force for consolidation and support in confused times, especially in view of the fact that it was not only the secular empire that had fallen apart. Thanks to a worrying tendency among the Byzantine Greeks, Christendom itself was now in danger of cracking up. A row was developing about the use of a single word that was to do more to change the course of history than perhaps any other. The word was *filioque*, Latin for "and from the Son." The reason anybody cared about this arcane Latin term was that its inclusion in a Christian creed indicated that the user followed the rule of Rome.

In essence, ever since the first century the official Christian line had been that the teacher and guide of the church was the Holy Spirit and that through baptism, its power entered and sanctified each individual in preparation for the day of resurrection. So the Spirit was essential to salvation. The problem arose because Saint Augustine had made it clear that while the Son was born of the Father, the Holy Spirit proceeded principally from the Father but *also* from the Son. Thus, the Christian creed referred to the Holy Spirit as coming from the Father, *filioque* ("and from the Son"). Unfortunately, this was not how the Byzantine Greeks saw things. For them, the Spirit came *only* from the Father, so they removed the word *filioque* from their creed.

It was for this reason that in A.D. 589 a great church council was convened in Toledo, Spain. At the instigation of the Visigothic king, Recared, the meeting formally anathematized any individual or body that did not include the word *filioque* in its creed. All books expressing any contrary belief were to be burned. This kind of support, from the most important secular authority in Europe, was good news for Rome. So the use of *filioque* spread in the West, where by 1013 its inclusion in liturgy was universal.

At that time, however, a council held by a bunch of nobodies in a tinpot Western town was of no consequence to the glitterati of the Byzantine Empire. Byzantium was the sophisticated center of the world. It had technology that the West couldn't even spell, let alone understand. The empire had a diplomatic corps *en poste* in countries as far away as India. The Byzantines possessed a professional bureaucracy, an organized system of finance and law, a navy, gold coinage and an educated, literate citizenry. The Roman pope's opposition to the way they said their creed was a matter of supreme indifference. *Filioque*, they claimed, was merely a "Latin" addition to the creed; this alone was justification for any Greek to ignore it.

This attitude seemed justifiable enough, while Byzantine fortunes were riding high and the pope was only some scruffy prelate out in the sticks. But things were soon to change. One day, at the Battle of Manzikert, in Armenia, the Byzantine army was unexpectedly destroyed by a bunch of Turkish freebooters on the rampage. They were led by a chief called Seljuk, and their mounted archers were invincible. The Byzantines panicked and appealed to the West for help. The papal response, in 1085, was the Crusade,[300] the first of several that would follow, as the Turks gradually increased their grip on the Balkans and came closer and closer to Constantinople itself.

300 89 64

By the fourteenth century the Turks were almost at the city gates, and the Byzantines were appealing to anybody who would listen: in Russia, Venice, France, England, Spain and, of course, Italy. In 1399 Manuel II, the emperor himself, was on the European-aid circuit. But all he got in response to his pleas was vague promises. By 1437 the new emperor, John VIII, was in Italy, together with five hundred "learned Greeks" for a meeting with the Catholics at Ferrara. Because of an outbreak of plague, the venue was changed to Florence,[301] where, on July 6, 1439, after much hag-   301   248   *211* gling, John finally accepted *filioque*. Agreement on the principle of Western aid was reached, but it was all too little, too late. Within four years Constantinople would fall to the Turks, and the *filioque* issue was history.

Most of the Greek delegates to the Council of Florence were now homeless, so many of them settled in Italy, especially in Venice, with which they had enjoyed strong economic and cultural ties for centuries. One of the immigrants, John Bessarion, bishop of Nicaea, had been so helpful to the Western side during the Florence talks that a grateful pope made him a cardinal of the Roman church, and later Latin patriarch of Constantinople. Pope Nicholas V might also have been attracted to Bessarion because of the library of more than five hundred books he had brought with him from Greece — which, in 1448, he donated to the basilica of San Marco in Venice. Nicholas was an ardent admirer of all things classical and had himself collected Greek books during his earlier travels.

Bessarion's house in Rome soon became an intellectual center for Greek translators and Italian humanists like Lorenzo Valla.[302] In   302   245   *209* one sense, the fall of Constantinople and the arrival of Greek immigrants — first in Venice and then throughout the rest of Italy — helped kick off the Renaissance. The Greek refugees gave their Italian hosts lessons in the Greek language and philosophy. For the first time, Italian intellectuals were getting the body of Greek thought straight from the horse's mouth. Most of the texts they were now studying had been lost to the West in the Dark Ages or had been available only through early medieval Arabic translations.

By 1469, once the first press had arrived in Venice, the Greek immigrants were also helping to publicize their country's literature in new, printed books. By 1496 the best-known printer in Venice was a Roman, Teobaldus (Aldus) Manutius, who took as his motto,

"Time Presses." If Aldus had one failing, it was his love for quality in an age when what mattered most was quantity. His first editions were criticized as being too lavish and expensive for ordinary scholars. It was this rapidly growing market for cheaper books that stimulated Aldus to produce the world's first pocket books, thanks to the invention by his Bolognese typemaker, Francesco Griffo, of a new typeface which became known as italic. The cursive font was based on a relatively new script, introduced early in the fifteenth century to cope with the increasing demand for speed in the papal chancellery.

The new, small print reduced the requirement for paper, which was the most expensive element in book production. The italic font cut the cost of books at a stroke of Griffo's punch-cutting chisel, which meant that Aldus could afford to be the first publisher of editions of over one thousand, a fivefold increase over previous print runs. This economy of scale contributed to further costcutting. Moreover, since italic was so small, the large paper sheet could be folded into eighths, and Aldus's new octavo editions fitted neatly into saddlebags. By the time of his death in 1515, Aldus had translated every major Greek classic. And because he was the first printer to use a logo, his little books were known as Aldine editions.

Printing had already spread across Europe like wildfire. Whereas in 1455 there had been no printed texts, by 1500, when Aldus was selling his new books, there were already twenty million books in thirty-five thousand editions (one book for every five members of the population). By 1500 there were two hundred printing presses, everywhere from Stockholm, Sweden, to Palermo, Sicily. And once pirates had begun producing bootleg copies of Aldus's cheap italic editions, books rapidly became the first massproduced consumer product, triggering the spread of the new capitalism that was emerging. Most of the books on sale were Bibles, as well as the greatest bestseller after the Bible, Thomas a Kempis's *Imitation of Christ*. Sales of Latin and Greek classics also grew fast, though it would be more than fifty years before their numbers approached those of liturgical texts.

By the sixteenth century the press had generated a new problem: there were just too many books. Printing had given a powerful incentive to the growing number of aristocrats and princes who collected books. There were enormous collections as early as the fifteenth century, when Matthias Corvinus, king of Poland, had amassed more than fifty thousand volumes. The first public library had opened in fifteenth-century Florence. There were also royal collections in France, England, Germany and Spain. But the difficulty with owning so many books was that there was as yet no proper method of putting them in order.

The man who found the solution was one of Queen Elizabeth of England's spies. His name was Thomas Bodley, and after a brief career as a Fellow of Merton College, Oxford (where he lectured in Greek), he toured Europe to improve his languages. On his return in 1580, he became a gentleman usher to the queen; and from 1585 he was employed by her on various secret missions, carrying personal letters in the queen's own hand to various European princes. These letters were so secret that Bodley was not even accompanied by a servant. By 1588 he was Elizabeth's permanent resident in Holland and a member of the English Council of State. Finally, in 1598 Bodley returned to England, where court intrigue soon became too much for him, and he took up the less-dangerous life of an academic. Anticipating this move, in 1597 he wrote to the Oxford vice-chancellor asking to restore the old university library, which had been destroyed during earlier "administrative reforms."

Bodley was determined that his new library would be the best ever, and to that end he used his many contacts to obtain books, promising in return that benefactors' names would appear emblazoned on the walls of the new library buildings. He also employed a bookseller to travel to France, Italy, Spain and the Frankfurt book fairs in search of texts. He instructed his librarian, Thomas James, to "take no riffe raffe bookes for such would prove a discredit to our Librarie." In 1601 he persuaded the official Stationers' Company to donate a copy of every book they published. In 1603, when the Bodleian Library was formally opened, there were books in thirty languages, the library was free for masters of the university to use six hours every day except Sundays and holy days, and King James had endowed it with the income from a farm in Berkshire and properties in London. The library's chief claim to fame, however, was that it boasted the first-ever general catalog, which set the style for all other libraries that followed.

But the main complaint, soon to be voiced about the Bodleian and other libraries, related to the matter of "riffe raffe bookes," by which Bodley had meant those written in vernacular English. The flood of texts grew, unabated, through the early seventeenth century, with books in every discipline — from mining to astronomy, animal husbandry to botany. And as the European economy boomed from the impact of expanding transoceanic trade and technology, it became more and more urgent to restructure the educational system to meet increasing demands for a curriculum and textbooks that would be more relevant to the needs of the new nation-states.

One of the first people to do something about this need was a Czech Protestant ex-priest, Jan Amos Komensky, better known by his Latin name of Comenius. In 1631, at the age of thirty-nine, he wrote the first language textbook that taught Latin by placing the words on the page, next to their Czech equivalents. Called *The Gate of Languages Unlocked*, the book was so successful that a version appeared in other European (and even Oriental) languages. This made his reputation among liberal thinkers, one of whom, the English Protestant Samuel Hartlib, invited him to England to establish a college of social reform. But in 1642 the English Civil War began, and plans for the college were shelved.

However, it may have been this link with England that brought Comenius his next offer. After he left for Holland, it seems that he was approached by John Winthrop, son of the Massachusetts Bay Colony governor, who asked him if he would accept the position of president of the new Harvard College. Although Comenius was attracted by Harvard's plans for teaching Native Americans and by the thought that New England would offer opportunities for social experiment, he refused the post. There seems to be some evidence, however, that his textbooks were in use at Harvard as early as 1650.

Meanwhile, Comenius continued to push for better forms of vocationally oriented education and instituted his new "pansophist" movement, dedicated to finding unity in all forms of human knowledge. His primary aim was to encourage the study of the sciences, both as the source of much new knowledge and as the best kind of preparation for the world of work. This Comenian view of vocational education was finally to come to flower all over Europe and America (ironically, because of the reason he himself had been obliged to leave England). The English Civil War, which temporar-

ily ended the monarchy, had brought Oliver Cromwell[303] to power
and established the republican Commonwealth of England.

303 156 *124*
303 208 *168*

In 1660 this social and political experiment ended with the restoration of the monarchy. This was accompanied by new legislation requiring all free-church Protestants (who had led the revolution that had brought Cromwell to power) to admit their defeat and to sign an oath of loyalty to the restored monarchy and its Anglo-Catholic church. Those who did not assent to swearing this oath were called "Dissenters," and their lives were made extremely difficult. The fate of the Dissenters was a perfect example of the adage, "Be nice to people on the way up, because you may need them on the way down." When the Protestant Republicans had first won the English Civil War, they had immediately set up a Committee for Plundered Ministers and started to eject those clergy judged to be "scandalous, insufficient and malignant." One Hugh Robinson, son of a bishop, was made to ride his horse facing backward all the way to jail. Churches were dismantled, the Book of Common Prayer was banned and all "sports" were forbidden on Sundays.

So when the Commonwealth failed and the monarchy was restored, there were many old scores to be settled, and the Dissenters now got more than they had given. Two acts of Parliament were passed in quick succession, setting up the Clarendon Code, and then further acts were passed, all designed to deprive Dissenters of the right to hold office in government or church and to prevent them from holding meetings. Municipal officials now had to be of the Anglican religion, and every minister had to swear to make no attempt to bring change to the Established Church. As a result of this code, nearly a thousand Dissenting ministers were ejected. In 1664 another new law prevented meetings of more than five people for any religious purposes other than those of the Anglican church. Three offenses would result in the condemned parties being transported to the colonies (but not to Puritan New England, where they might be welcomed with open arms). In 1665 Dissenting ministers or teachers were not permitted within five miles of a borough.

The Quakers came off worst, being fined for keeping schools, for not removing their hats, for not paying tithes and for refusing to muster arms. Fines were often ruinous; when an offender had no money, his or her possessions were sequestered. Quaker meetings were violently broken up by constables with drawn swords. Some families were taken from their beds at night and made to run along-side the officers' horses for up to twenty miles to prison. Once in the disease-ridden prisons of the time, many Quakers died. Because of all this harassment, large numbers of Dissenters left the country for America or the Netherlands. Those who remained had little choice but to try and build a life in commerce and industry, since this was the only activity open to them without breaking the law. So throughout the country, by the early eighteenth century the majority of industrialists were Dissenters, and here and there they formed networks of liberal thinkers brought together by persecution. Thomas Newcomen, the inventor of the steam engine (improved by James Watt) first set up in business thanks to a fellow Baptist backer. In the English Midlands, the Quaker Charles Lloyd married into two other Dissenter families, both of which were iron-makers, thus laying the financial foundations for his famous bank.

In the long run, though, it was the fact that Dissenters were denied all other forms of career that brought success to the men whose names now read like a roll call of famous inventors: James Watt,[304] Matthew Boulton,[305] Thomas Newcomen, Joseph Priestley,[306] Samuel Barclay, Josiah Wedgwood[307] and others developed the technology, as well as the scientific and financial institutions, that would make possible the Industrial Revolution. And since Dissenters were also not permitted to send their children to university, and because of the demands of the industrial and commercial worlds in which they were constrained to work, first the Presbyterians and then others set up schools of their own. These schools were the first to teach their pupils subjects that were relevant to the industrial world in which they would have to make their living.

The first Dissenter academies,[308] established by twenty ejected ministers, had aimed primarily to produce educated men for the pulpit. But by 1690 there were twenty-three such schools in England, and at least half of them allowed admission to boys who were not intending to take the cloth. Already by this time the curriculum was vocational in content and strongly influenced by Comenius's educational ideas, as well as his approach to textbook design. The schools taught science with the aid of the latest air pumps, thermometers and math instruments. They taught commercially important foreign languages like French. By the end of the eighteenth century, one of the best academies was in the northeast of England,

304 219 *184*
304 270 *227*
305 221 *184*
306   5 *14*
306 161 *128*
306 282 *245*
307 223 *185*
307 293 *260*

308 162 *128*

in a small town called Kendal. The school was a Quaker foundation, financially well endowed and possessing a large library, a telescope and a microscope. Kendal was on one of the main roads between London and Scotland, so it was also well served by passing lecturers, like the eminent natural philosopher Joseph Banks.

One of the assistant masters at the school began to give public lectures on mechanics, optics, astronomy and the use of globes; but because the response was poor, he left the area and eventually settled in Manchester. There, in 1800, he set up a highly successful mathematical academy and, with the spare time his income now afforded him, took an interest in the weather. For fifty-seven years he took meteorological observations every day, and on the day he died his diary simply read: "Little rain today." His name was John Dalton, and he was profoundly impressed by the work a German had done on infinitely small matter.

The German in question, Gottfried Leibniz,[309] was, for much of his life, one of those people who never complete anything they start. He was a mathematician who made a wooden calculating machine which he couldn't get to work, so he was unable to complete his grand scheme to reduce all thoughts to numbers. According to Leibniz, this would have made it possible for people to arrive at solutions to problems by simply combining the necessary thought-numbers. Leibniz also failed to get a giant mine-drainage project started. He failed to turn all words into a collection of universal signs, so there would no longer be foreign languages. He contracted to write a complete history of the dukes of Brunswick (he was their family librarian) but never finished it. And he ran out of money.

However, Leibniz succeeded in one thing that changed the world. He worked out the kind of math needed to calculate the movement of the planets. This kind of calculation involved being able to work out the rate of change of acceleration or deceleration in the velocity of a planet, at any instant. The math Leibniz developed in 1675 to do this became known as infinitesimal calculus. For Leibniz, however, the mathematical uses of his calculus were not as important as the philosophical applications. It may have been his visit to Holland that same year (when he met Anton van Leeuwenhoek,[310] who had recently made his amazing discovery of microscopic living organisms) that turned Leibniz's mind to other infinitesimal matters, which would in turn later inspire Dalton.

The problem at the time was matter. What is it, and why is a

309 82 55
309 231 194
309 253 214

310 83 55
310 254 214

pile of sand made up of many tiny pieces, when a stone is apparently only a single piece? How far can matter be broken down? Theories existed regarding the existence of infinitesimally small pieces of matter, but nobody was able to show how they compose wholes. Leibniz suggested that no material object is so small that it cannot be subdivided. At some point, however, there has to be an infinitely small "essence of matter," which has to be a point composed solely of energy. These point-particles would be the fundamental elements of all existence. But how do these energy particles turn into matter? Leibniz invented entities called monads, describing them as the "requisites" of matter that hold together the point-particles and in this way permit matter to compose. The smallest group held together in this way would, therefore, be the smallest version possible of any material.

It was this thought that drove Dalton in his study of the weather. As part of his meteorological obsession Dalton built barometers, thermometers and rain gauges, so as to measure mists, rainfall, clouds and air pressure. Above all, Dalton was interested in humidity and how water vapor entered and left the atmosphere. Nobody at the time knew how air absorbed water. The prevalent theory was that there was some kind of chemical attraction at work. Dalton was convinced it was related in some way to air pressure. Pressure decreases with altitude, thereby indicating that gravity ought to keep most air particles close to the ground — which means they have to have weight.

Then Dalton discovered that air has four times more nitrogen than oxygen, so he began to investigate how gases combine. He noticed that after a number of gases are mixed, there is always some gas left that has not combined with the others. He then found that oxygen and hydrogen always mix in the ratio of 8:1. Other gases form similarly regular combinations. Is this related to their weights? When he tried forcing gases into water with pressure, some gases were readily absorbed but some were not. Does this also have something to do with their weight? And all gases always behave in their own, idiosyncratic way with water, seeming to prove that each individual gas is always the same. Does this mean a gas always has the same weight?

In 1803 Dalton published a paper on gas absorption by water, in which he explained that the absorption depends on the weight of the gas particles. He also added a list of these weights. In 1805 he produced the first proper table of weights, and in 1807 described to an audience in Edinburgh "a new view of the first principles or elements of bodies and their combinations." In the talk he claimed that this view would "produce the most important changes in the system of chemistry, and reduce the whole to a science of great simplicity, and intelligible to the meanest understanding." With this modest remark, Dalton set up modern chemistry and introduced the world to chemical atomic theory, by which compounds form according to the relative weights of their individual atomic components.

It took a Swedish hypochondriac with a taste for women, gourmet food and spas to make Dalton famous. In the year of Dalton's announcement Jöns Jakob Berzelius had just been appointed professor of medicine and pharmacy at the Stockholm Medical College (later known as the Karolinksa Institute). Here in his lab, Berzelius devoted himself to learning all that he could about chemistry. Thanks to a rich wife, Berzelius was able to devote himself to his pleasures, which included travel (he kept entertaining journals full of details on women's shapes), eating (a forty-course meal, on one French occasion), drinking mineral waters (everywhere, for his imaginary illnesses) and using his blowpipe. This was a common instrument of the day, mainly used to raise the temperature of a flame as high as 1,500°C (2,702°F) in order to analyze powdered minerals. Berzelius was very good at this and would often identify the elements in his hosts' rock collections (he did so for Goethe) in return for board and lodging.

Berzelius was much taken with Dalton's atomic-weights table and, in 1818, succeeded in identifying the weights of forty-five out of the forty-nine known elements, as well as listing over two thousand compounds. It was the sheer mind-numbing complexity of this work that convinced him that some better way had to be found to rationalize chemical notation. So he came up with the symbology used ever since. He identified each element by the first letter of its Latin name, unless there was more than one beginning with the same letter, in which case he also used the second letter of the Latin name. In the case of compounds, small subscript figures were attached to indicate proportions. Thanks to Berzelius, modern chemists write "sulfuric acid" (a combination of hydrogen, sulfur and oxygen) as H_2SO_4.

Berzelius did one other thing to endear himself to posterity. He identified the composition of a mysterious stone found near an iron mine in Sweden. It was a kind of tungsten, and Berzelius (and his rich host at the time) found that it was mixed with a new element. But instead of naming it Swedonium or Berzelium, as might have been expected, he named it cerium,[311] after the new asteroid which had recently been discovered (and then lost). The strangest thing about this unusual astronomical event was that the discoverer had known where to look in the first place. Back in 1772 the director of the Berlin observatory, Johann Bode, had spread the word about how there appeared to be a mathematical relationship involved in the distances separating the planets. Planets were found at regular intervals from the Sun, based on the following series of numbers: 0, 3, 6, 12, 24, 48, 96, 192. Each planetary interval was twice the previous value in the basic series, plus 4. This sum correctly predicted the position of all the planets except Neptune. But it also revealed that there should be a planet at the "24 + 4" position. But so far, none had been found at that position.

It was this mystery that planet astronomers were searching for when Ceres was discovered by Giuseppi Piazzi, at the Observatory in Palermo, Sicily, in 1801. Piazzi excitedly took three observations of the new planet and then fell ill. After he had recovered, the weather was too cloudy to observe for several weeks, but when it cleared, Piazzi looked again. To his chagrin, Ceres was gone, and there seemed no way to rediscover it. He had watched it through only about 9° of its orbit, which didn't provide enough data to extrapolate its present position. The news of the loss was a major blow to the world of astronomy. So when a young mathematical genius in Germany announced that he had developed a method by which Ceres could again be found, everybody was delighted but skeptical. However, sure enough, Karl Friedrich Gauss had found a way to calculate an orbit from only three observations; a year later, to the day, Ceres was rediscovered, just where Gauss had said it would be. The event made him an instant world celebrity and earned him the post of astronomer at the University of Göttingen, where they have preserved his brain to this day.

Gauss was a man of many interests. He invented the telegraph before Morse and dabbled in a strange, ancient language that was becoming extremely popular among Romantically inclined Germans. The language was Sanskrit,[312] and it had first come to the 312 278 *239* attention of scholars in the West thanks to the efforts of a Welsh judge, Sir William Jones, who was working in India. Jones was a language prodigy and had learned Hebrew, Latin, Greek, French, Italian, Arabic, Portuguese and Spanish. He had also published a Persian grammar before arriving in Calcutta in 1783. There he discovered the delights of Sanskrit. Jones was a religious man and was convinced that one day proof would be found that the ancient stories in the Bible were all true. So Sanskrit fired his imagination because he thought it might turn out to have been the Edenic tongue. By 1784 he was well on the way to mastering the language.

In 1786 Jones electrified the Western world with his announcement that he thought Sanskrit was the language from which all European languages were descended. He himself traced the links with Greek and Latin, Celtic, Armenian and Albanian. Sanskrit rapidly became the fashionable language of European scholarship once Jones had produced a grammar. It was at this point that German scholars became interested, and one of them introduced Gauss to the subject. Part of the reason Germans in particular were so interested in the ancient past was that at the end of the nineteenth century, the spirit of Romanticism and revolution (French and American) was in the air, and these had kindled the fires of nationalism.

Thanks to Romanticism, the mechanistic orderliness of the ancien régime worldview was by now giving way to individualism and the new glorification of "feeling." Part of this trend was a Romantic revival of interest in the medieval past, which gave rise to a desire to find the roots of each culture's own identity. The German thinker Johann Herder conjured up a mystic, ancient tribal German *volk*, which was neither a political nor economic entity, but rather some kind of fundamental, cultural archetype from which the German race had sprung and which gave it a uniquely German character. This Teutonic identity, he warned, was at danger from the insidious "international" character of French Enlightenment thinking. In an eerie forshadowing of eugenics and racial purity, Herder called on peoples everywhere to shake off the French philosophical yoke and to seek their own national cultural heritage as expressed in their folklore.

By 1806 all this had been given added urgency by the Napole-

313 212 *172* onic defeat of the Prussians at the Battle of Jena.[313] German Romantics looking to restore national pride found in Sanskrit an ancient Indo-European origin from which their *volk* had originally sprung. It was this imagined Aryan past on which Wagner would draw for his heroic German operas and that would inspire mad King Ludwig of Bavaria to build imitation medieval castles like Neuschwanstein, complete with electric lights. Aryanism also gave a philologist whose Sanskrit work had interested Gauss the idea of turning linguistics into a science.

His name was Jacob Grimm, and for a short time he had been at the University of Göttingen together with Gauss. Grimm made a major contribution to linguistics when, between 1822 and 1837, he joined the Indo-European bandwagon and published four massive volumes of analyses of the sound changes he had traced in the development of European languages from their ancient roots. He concentrated particularly on sounds that seemed to link the Germanic language family. For instance, the Latin *p* (as in *pater*) became an *f* in English and German ("father," *fater*); the Latin *d* became *t* (from *duo* to "two," *zwei*); Sanskrit *dh* became Greek *ph*, Latin *f* and Germanic *b* or *v*.

Jacob also had a brother, Wilhelm, from whom he was never parted and who shared Jacob's obsession with the Germanic past. Both brothers saw in language and folklore a chance to express the fundamental bond that unified all Germans. Jacob said, "All my works relate to the Fatherland, from whose soil they derive their strength." Sometime around 1807 both men started collecting folk material by inviting into their home people who could tell old tales. They collected stories from gypsies, peasant women, shepherds, wagoners, vagrants, grandmothers and children. When they published their final one, in 1852, they had written a total of 211 stories, many of which came from as far afield as Persia, Sweden and India.

The *Grimm Brothers' Fairy Tales* would enchant generations of children. In fact, they were originally inspired by Herder's call for stories of German culture for the children of the future. But the original *Tales* were different from the toned-down texts of later editions and very different indeed from the sanitized versions one day used by Disney. The first versions of the *Tales* were full of primal violence: Rapunzel is made pregnant by the Prince; Cinderella's ugly sisters have their eyes pecked out; the old woman cuts her daughter's head off; Hansel and Gretel cook the witch in the oven; the evil woman in Snow White is forced to dance in red-hot shoes; and Sleeping Beauty is a tale of necrophilia.

The *Tales* aroused immediate public interest because of the way the elements of many of the stories were so similar, no matter from which culture they originated. In Serbia, England or Norway, the same episodes reappeared again and again, expressing the same values. Virtue was always rewarded, and violence and fear were always glorified. Strangers were regarded with contempt. Order, discipline and authoritarianism were always held in high regard. Family unity was challenged by poverty or the complications of inheritance. Daily life was frequently interrupted by cruelty, violence or torture. The prime virtue was courage.

Little Red Riding Hood became the subject of scholarly study in 1865 when an Englishman named Edward Tylor found similarities between the story of the wolf swallowing the girl (and her grandmother) and the ancient Nordic myth of Skoll, the sun-devouring wolf. Tylor saw the tales as even more ancient than the Grimms had thought them to be. For Tylor, these were primeval mythological explanations for natural processes like new growth every spring, where the sun went at night, how the clouds swallowed up the earth and so on. Little Red Riding Hood herself was descended from mythic representations of dawn or the Moon.

In 1865 Tylor expressed these ideas in *Early History of Mankind*, the first proper work of modern anthropology.[314] In his second, even more successful book, published in 1871, his title coined two basic modern anthropological terms: *Primitive Culture*. Tylor drew on his travels and his study of archeology, linguistics, history, geography, paleontology and, above all, folklore, to produce a grand theory of social development. Tales like those collected by the Grimms told the same stories because all human groups shared a common developmental history. All modern societies had originated as primitive, savage groups and had passed through the same stages on the way to modern social forms. During this progression they had developed the same stages of knowledge and tool-use — from shaped pebbles to fire to bow. They had also asked the same fundamental questions about the universe and had constructed myths to answer them. Tylor clinched his argument with examples of what

314   154   *124*

he called "survivals." These were ancient customs and habits that had survived through eons, long after their original purpose had been lost and forgotten. They included examples like throwing salt over the shoulder, the bridal ring, the aversion to walking under ladders, the costumes of all religions, the kiss under the mistletoe and many others. In our superstitions, said Tylor, we reveal our common ancestry.

This final chapter begins with the attempt by the Catholic church to hold things together after the fall of Rome, by the enforcement of conformity (through the use of *filioque*). It ends in the present day, after the fall of another world order, with the emergence of social anthropology, a science that persuades us we will survive the chaos of the post–Cold War era only by celebrating the differences that exist among us.

This book attempts to draw the same conclusions about knowledge as Edward Tylor did about humanity. It is interdependent. Nothing on the great web of change exists in isolation. I try to show this by the way, in many cases, an episode in the timeline of one chapter can as easily also lead directly to an event on a totally different timeline. I hope the reader will try the exercise at least once, to get a feel for the crazy way the pinball of change works its magic, bouncing here and there across time and space. There is no single, correct pathway on the web, or in life. Mistrust anybody who tells you so.

One final twist is one you might have already suspected. This is not the end of the book. As with the web, there is no end. This twentieth chapter finishes with the anthropologists' discovery of cultural diversity. One of the most powerful ways this diversity has been expressed is in the way, through history, people have adorned themselves. Some of the most distinctive forms of adornment have been the many different styles with which women in different societies have dressed their hair.

In our own time, this has been manifested in the extraordinary growth of the multibillion-dollar beauty industry. With which this story continues, at the beginning of the book. . . .

Select Bibliography

Abbott, George C. *Sugar.* London: Routledge, 1990.

Ackerknecht, Erwin H. *Rudolph Virchow: Doctor, Statesman and Anthropologist.* Madison: Wisconsin University Press, 1953.

Aiton, E. J. *Leibniz.* Bristol, England: Adam Hilger, 1985.

Allan, D. G. C., and R. E. Schofield. *Stephen Hales.* London: Scholar Press, 1980.

Allen, N. *David Dale, Robert Owen and the Story of New Lanark.* Edinburgh: Mowbray House Press, 1986.

Allison, H. E. *Benedict de Spinoza: An Introduction.* New Haven: Yale University Press, 1987.

Anand, R. P. *The Origin and Development of the Law of the Sea.* The Hague: Martinus Nijhoff, 1983.

Andrews, Henry N. *The Fossil Hunters.* Ithaca: Cornell University Press, 1980.

Baldwin, E. *Gowland Hopkins.* London: Van den Berghs Ltd., 1962.

Barakat, R. A. *The Cistercian Sign Language.* Kalamazoo, Mich.: Cistercian Publications, 1975.

Barraclough, Kenneth C. *Benjamin Huntsman.* Sheffield, England: Sheffield City Libraries, 1976.

Barthorp, M. *Napoleon's Egyptian Campaigns, 1798–1801.* London: Osprey Publishing, 1978.

Bassett, John M. *Samuel Cunard.* Post Mills, Vt.: Fitzhenry & Whiteside, 1976.

Baxter, J. P. *The Introduction of the Ironclad Warship.* Cambridge: Harvard University Press, 1933.

Beard, G. *The Work of Robert Adam.* Edinburgh: J. Bartholomew & Sons, 1978.

Bell, A. E. *Christiaan Huygens and the Development of Science in the Seventeenth Century.* London: Edward Arnold, 1947.

Bleich, A. R. *The Story of X-Rays.* New York: Dover Publications, 1960.

Blunt, Wilfrid. *The Ark in the Park.* London: Hamish Hamilton, 1976.

Boring, Edwin B. *A History of Experimental Psychology.* New York: Century, 1929.

Bourne, J., et al. *Lacquer: An International History.* Marlborough, England: Crowood Press, 1984.

Boxer, C. R. *The Portuguese Seaborne Empire, 1415–1825.* London: Hutchinson, 1969.

Brandi, Karl. *The Emperor Charles V.* Brighton, England: Harvester Press, 1980.

Brandon, R. *Singer and the Sewing Machine.* London: Barrie & Jenkins, 1977.

Brown, Chandos Michael. *Benjamin Silliman: A Life in the Young Republic.* Princeton, N.J.: Princeton University Press, 1989.

Brown, Lloyd A. *The Story of Maps.* London: Cresset Press, 1951.

Brozek, Josef, and Horst Gundlach. *G. T. Fechner and Psychology.* Passau, Germany: Passavia, 1988.

Bruton, Eric. *The History of Clocks and Watches.* London: Orbis, 1979.

Buhler, W. K. *Gauss. A Biographical Study.* Berlin: Springer, 1981.

Burton, Anthony. *Josiah Wedgwood.* London: André Deutsch, 1976.

———. *The Rise and Fall of King Cotton.* London: André Deutsch, 1984.

Campbell, George F. *China Tea Clippers.* London: Adlard Coles Nautical, 1974.

Carrier, Willis H., ed. *Modern Air-Conditioning, Heating and Ventilating.* Aulander, N.C.: Pitman Publishing, 1940.

Cassirer, Ernst. *Kant's Life and Thought.* New Haven: Yale University Press, 1981.

Chandler, Alfred D. *The Visible Hand. The Managerial Revolution in American Business.* Cambridge: Harvard University Press, 1977.

Cheney, Margaret. *Tesla.* Englewood Cliffs, N.J.: Prentice-Hall, 1981.

Clark, R. W. *Einstein.* London: Hodder & Stoughton, 1979.

Cohen, Robert S., and R. J. Seeger, eds. *Ernst Mach, Physicist and Philosopher.* Dordrecht, Netherlands: D. Reidel Publishing Co., 1970.

Cole, Charles W. *Colbert and a Century of French Mercantilism.* London: Frank Cass & Co., 1964.

Coleman, C. B. *The Treatise of Lorenzo Valla on the Donation of Constantine.* New Haven: Yale University Press, 1922.

Collis, Maurice. *Raffles.* London: Century, 1988.

Davidson, H. *Pascal.* Boston: Twayne Publishers, 1987.

Dawson, T. R., ed. *History of the Rubber Industry.* Cambridge, England: W. Heffer & Sons, 1953.

de Jong, Cornelis. *A Short History of Old Dutch Whaling.* Pretoria: 1978.

De Sola Pool, Ithiel, ed. *The Social Impact of the Telephone.* Cambridge: MIT Press, 1977.

Demus, Otto. *Byzantine Mosaic Decoration.* London: Routledge & Kegan Paul, 1976.

Desmond, A., and J. Moore. *Darwin.* Harmondsworth, England: Penguin, 1991.

Deuel, Leo, ed. *Memoirs of Heinrich Schliemann.* London: Hutchinson, 1978.

Dickinson, H. W. *John Wilkinson, Ironmaster.* Ulverston, England: Hume Kitchin, 1914.

Dorman, C. C. *The Stephensons and Steam Railways.* London: Priory Press Ltd., 1975.

Downs, Robert B. *Heinrich Pestalozzi, Father of Modern Pedagogy.* Boston: Twayne Publishers, 1975.

Duffy, C. *Frederick the Great.* London: Routledge & Kegan Paul, 1985.

Duncum, Barbara. *The Development of Inhalation Anaesthesia.* Oxford: Oxford University Press, 1947.

Dunkel, H. B. *Herbart and Herbartianism: An Educational Ghost Story.* Chicago: University of Chicago, 1970.

Eder, Josef Maria. *History of Photography.* New York: Dover Publications, 1978.

Ehrenberg, Richard. *Capital and Finance in the Age of Renaissance.* New York: Augustus Kelley, 1963.

Ewald, Peter. *Fifty Years of X-Ray Diffraction.* Utrecht, Netherlands: International Union of Crystallography, 1962.

Fehervari, G. *Islamic Pottery.* London: Faber & Faber, 1973.

Fest, Joachim C. *Hitler.* Harmondsworth, England: Penguin, 1977.

Ford, Brian J. *The Leeuwenhoek Legacy.* London: Biopress, 1991.

Forrest, D. W. *Francis Galton: The Life and Work of a Victorian Genius.* London: Elek, 1974.

Fox-Genovese, Elizabeth. *The Origins of Physiocracy, Economic Revolution and Social Order.* Ithaca: Cornell University Press, 1976.

Gannon, Jack R. *Deaf Heritage.* Silver Spring, Md.: National Association for the Deaf, 1982.

Gibbs, F. W. *Joseph Priestley: Adventurer in Science and Champion of Truth.* London: Thomas Nelson & Sons, 1965.

Gibson, A. H. *Osborne Reynolds.* London: Longmans, Green & Co., 1946.

Goldensohn, B. *The Marrano.* Orono: University of Maine Press, 1988.

Green, C. M. *Eli Whitney and the Birth of American Technology.* Boston: Little, Brown, 1956.

Greenberg, M. *British Trade and the Opening of China 1800–1842.* Cambridge: Cambridge University Press, 1951.

Gunston, D. *Guglielmo Marconi.* Geneva: Edito-Service, 1970.

Hall, G. S. *Founder of Modern Psychology.* New York: Appleton & Co., 1912.

Hammitszchen, H. *Zen and the Art of Tea Ceremony.* Harmondsworth, England: Penguin, 1979.

Harding, H. *Tunnelling History.* Toronto: Golder, 1981.

Hardy, Robert. *Longbow: A Social and Military History.* Sparkford, England: Patrick Stephens Ltd., 1992.

Herrmann, D. B. *The History of Astronomy from Herschel to Hertzsprung.* Cambridge: Cambridge University Press, 1973.

Hey, Colin G. *Rowland Hill: Victorian Genius and Benefactor.* London: Quiller Press, 1989.

Higham, Norman. *A Very Scientific Gentleman: The Major Achievements of Henry Clifton Sorby.* Oxford, England: Pergamon Press, 1963.

Hill, Christopher. *God's Englishman*. Harmondsworth, England: Penguin, 1972.

Hobden, Heather, and Mervyn Hobden. *John Harrison and the Problem of Longitude*. Lincoln, England: Cosmic Elk, 1989.

Homer, W. I. *Seurat and the Science of Painting*. Cambridge: MIT Press, 1964.

Honour, Hugh. *Neoclassicism*. Harmondsworth, England: Penguin, 1968.

———. *Romanticism, Style and Civilisation*. London: Allen Lane, 1979.

Hounshell, David. *From the American System to Mass Production, 1800–1932*. Baltimore: Johns Hopkins University Press, 1984.

Hughes, T. P. *Elmer Sperry, Inventor and Engineer*. Baltimore: Johns Hopkins University Press, 1971.

Hunter, James M. *Perspective on Ratzel's Political Geography*. Lanham, Md.: University Press of America, 1983.

Inglis, Brian. *The Opium War*. London: Hodder & Stoughton, 1976.

Jensen, Oliver. *The American Heritage History of Railroads in America*. New York: Bonanza Books, 1987.

Jenyns, Soame. *Ming Pottery and Porcelain*. London: Faber & Faber, 1988.

Kellner, Charlotte. *Alexander von Humboldt*. Oxford: Oxford University Press, 1963.

Kind, Stuart. *The Scientific Investigation of Crime*. Harrogate, England: Forensic Science Services, 1987.

King, H. C. *History of the Telescope*. London: Charles Griffin & Co., 1955.

Koenigsberger, Leo. *Herman von Helmholtz*. Oxford, England: Clarendon Press, 1906.

Kurylo, F. *Ferdinand Braun*. Cambridge: MIT Press, 1981.

Leitch, D., and A. Williamson. *The Dalton Tradition*. Manchester, England: John Rylands University Library, 1991.

Leopold, Joan. *E. B. Tylor and the Making of Primitive Culture*. Berlin: Reimer, 1980.

Lewis, M. A. *The Spanish Armada*. London: B. T. Batsford Ltd., 1960.

Lovejoy, Arthur O. *The Great Chain of Being*. Cambridge: Harvard University Press, 1936.

Macfarlane, Gwyn. *Alexander Fleming: The Man and the Myth*. Oxford: Oxford University Press, 1984.

Macintyre, Donald. *The Privateers*. London: Elek, 1975.

MacKinnon, Neil. *This Unfriendly Soil. The Loyalist Experience in Nova Scotia, 1783–1791*. Kingston, Canada: Queen's University Press, 1986.

Marshall, P. H. *William Godwin*. New Haven: Yale University Press, 1984.

Marx, R. *Port Royal Rediscovered*. London: New English Library, 1973.

McCord, Norman. *The Anti–Corn Law League*. London: Unwin University Books, 1962.

McGlathery, J., ed. *The Brothers Grimm and Folktale*. Champaign: University of Illinois Press, 1988.

Melhado, Evan M. *Jacob Berzelius: The Emergence of His Chemical System*. Stockholm: Almqvist & Wiksell International, 1980.

Mellor, Anne K. *Mary Shelley.* London: Routledge, 1988.

Merrill, G. P. *The First One Hundred Years of American Geology.* New Haven, Conn.: Philip Hamilton McMillan Memorial Publishing Fund, 1924.

Middleton, W. E. K. *The History of the Barometer.* Baltimore: Johns Hopkins University Press, 1964.

Mitchell, D. *Pirates.* London: Thames & Hudson, 1976.

Moffit-Watts, P. *Nicholas Cusanus.* Leiden: E. J. Brill, 1982.

Moore, Patrick. *Patrick Moore's History of Astronomy.* London: Macmillan & Co., 1983.

Morris, R. J. *Cholera.* London: Croom Helm Ltd., 1976.

Moulton, F. R. *Liebig and After Liebig.* Pennsylvania: Science Press Co., 1942.

Muirhead, J. P. *Life of James Watt.* London: Archival Facsimiles, 1987.

Murray, C. H. *Apollo: The Race to the Moon.* London: Secker & Warburg, 1989.

Nicholas, David. *Medieval Flanders.* Harlow, England: Longman, 1992.

Norris, J. D. *Advertising and the Transformation of American Society.* New York: Greenwood Press, 1990.

Osborn, Frederick M. *The Story of the Mushets.* London: Thomas Nelson & Sons, 1952.

Paley, Edmund, ed. *An Account of the Life and Writings of William Paley.* Farnborough, England: Gregg International Publications, 1970.

Palmer, R. *The Water Closet.* Newton Abbot, England: David & Charles, 1973.

Pierson, Peter. *Philip II of Spain.* London: Thames & Hudson, 1975.

Piper, R. *The Story of Computers.* London: Hodder & Stoughton, 1977.

Pratt, Fletcher. *Secret and Urgent: The Story of Codes and Ciphers.* Indianapolis: Bobbs-Merrill Co., 1939.

Price, M. J., ed. *Coins: An Illustrated Survey.* London: Hamlyn, 1980.

Raven, Charles E. *John Ray Naturalist: His Life and Works.* Cambridge: Cambridge University Press, 1942.

Reader, W. J. *Macadam.* London: Heinemann, 1980.

Rose, W. Lee. *Rehearsal for Reconstruction: The Port Royal Experiment.* Indianapolis: Bobbs-Merrill Co., 1964.

Schuyler, Hamilton. *The Roeblings.* Princeton, N.J.: Princeton University Press, 1931.

Scott, Samuel F. *French Aid to the American Revolution.* Ann Arbor, Mich.: William L. Clements Library, 1976.

Shockley, W. *Electronics and Holes in Semiconductors.* New York: Van Nostrand Co., 1950.

Skempton, A. W. *John Smeaton F.R.S.* London: Thomas Telford Ltd., 1981.

Snyder, L. L. *The Roots of German Nationalism.* Bloomington: University of Indiana Press, 1978.

Sonnino, Paul, ed. *The Reign of Louis XIV.* London: Humanities Press International, 1991.

Speed, P. F. *The Potato Famine and the Irish Emigrants.* London: Longman, 1976.

Stamp, T., and C. Stamp. *James Cook, Maritime Scientist.* Whitby, England: Caedmon of Whitby Press, 1978.

Sutter, J. A. *New Helvetia Diary.* California: The Crabhorn Press, 1939.

Tannahill, Reay. *Food in History.* London: Penguin, 1988.

Thomas, D. O. *Richard Price, 1723–1791.* Cardiff: University of Wales, 1976.

Thompson, F. M. L. *The Rise of Suburbia.* Leicester, England: Leicester University Press, 1982.

von Poelnitz, G. *Baron Anton Fugger.* Tübingen, Germany: Studien zur Fugger-geschichte, 1958–86.

Wason, K. *Delftware.* London: Thames & Hudson, 1980.

Watts, Michael R. *The Dissenters.* Oxford, England: Clarendon Press, 1978.

Webster, John C. *The Life of J. F. W. des Barres.* New Brunswick, Canada: Shediac, 1933.

Weinberg, Steve. *The First 3 Minutes.* London: André Deutsch, 1977.

Williams, L. P. *Michael Faraday: A Biography.* London: Chapman & Hall, 1965.

Wilton-Ely, J. *The Mind and Art of G. B. Piranesi.* London: Thames & Hudson, 1988.

Wizniter, Arnold. *Jews in Colonial Brazil.* New York: Columbia University Press, 1960.

Wright, Lawrence. *Clean and Decent.* London: Routledge & Kegan Paul, 1960.

Wyld, Lionel D. *Low Bridge! Folklore and the Erie Canal.* Syracuse, N.Y.: Syracuse University Press, 1962.

Index